NIST Special Publication 800-165

Computer Security Division

2012 Annual Report

Patrick O'Reilly, Editor
Computer Security Division
Information Technology Laboratory
National Institute of Standards and Technology

Lorie Richards
Facilities Services Division
Creative and Printing Service

June 2013

U.S. Department of Commerce
Penny S. Pritzker, Secretary

National Institute of Standards and Technology
Patrick D. Gallagher, Under Secretary of Commerce for Standards and Technology and Director

Disclaimer: Any mention of commercial products is for information only; it does not imply NIST recommendation or endorsement, nor does it imply that the products mentioned are necessarily the best available for the purpose.

Table of Contents iii

Welcome Letter 1

Division Organization 2

Intro to CSD's 5 Groups 3

   Cryptographic Technology Group 3

   Security Components and Mechanisms Group 4

   Secure Systems and Applications Group 4

   Security Outreach and Integration Group 5

   Security Testing, Validation, and Measurement Group ... 6

The Computer Security Division Implements
the Federal Information Security Management Act
of 2002 7

CSD's Programs/Projects for FY2012 8

   CSD Work in National and International Standards 8

   Identity Management Standards within INCITS B10
   and ISO JTC1/SC 17 11

   Federal Information Security Management Act (FISMA)
   Implementation Project 12

   Biometric Standards and Associated Conformity
   Assessment Testing Tools 15

   Cybersecurity of Cyber Physical Systems 18

   Cybersecurity Research & Development (R&D)......... 18

   Security Aspects of Electronic Voting 20

   Health Information Technology Security 21

   ICT Supply Chain Risk Management...................... 21

   Nationwide Public Safety Broadband Network (NPSBN)
   Security 23

   Smart Grid Cybersecurity................................ 23

   Cybersecurity Awareness, Training, Education, and
   Outreach... 26

   Cryptographic Technology 34

   Validation Programs...................................... 47

   Identity Management 54

   Research in Emerging Technologies 55

   Strengthening Internet Security 59

   Access Control and Privilege Management.............. 60

   Advanced Security Testing and Measurements 62

Honors and Awards 72

2012 Publications Released and Abstracts......... 73

   Abstracts for Publications Released in FY2012 75

      Federal Information Processing Standards (FIPS) 75

      Special Publications......................................76

      NIST Interagency Reports 83

Ways to Engage with Our Division and NIST 89

   Guest Research Internships at NIST 89

   Details at NIST for Government or Military Personnel ...89

   Federal Computer Security Program Managers' Forum
   (FCSPM) 89

   Security Research 89

   Funding Opportunities at NIST...........................89

Acknowledgements 90

With the continued proliferation of information, the explosion of devices connecting to the expanding communication infrastructure and the evolving threat environment, the need for cybersecurity standards and best practices that address interoperability, usability and privacy continues to be critical for the Nation. The Computer Security Division (CSD), a component of the Information Technology Laboratory at the National Institute of Standards and Technology (NIST) is responsible for developing standards, guidelines, tests, and metrics for the protection of non-national security federal information and communication infrastructure. These standards, guidelines, tests, and metrics are also important resources for the private sector.

In 2012, CSD aligned its resources to enable greater development and application of practical, innovative security technologies and methodologies, and to enhance our ability to address current and future computer and information security challenges in support of critical national and international priorities.

CSD extended its research and development agenda for high-quality, cost-effective security and privacy mechanisms to foster improved information security across the federal government and the global information security community. In 2012, NIST concluded the five-year SHA-3 Cryptographic Hash Algorithm Competition with the selection of KECCAK for standardization and worldwide adoption. The selection of this cryptographic hash algorithm, an indispensable component for the information and communication systems that support commerce in the modern era, confirmed NIST's well-respected and trusted technical authority in this field.

A strong partnership with industry is vital to the success of our technical programs. In 2012, CSD engaged the international standards community on emerging technical standards to advance continuous monitoring capabilities and to further interoperability with commercially available tools. We also devised new security testing methodologies for smart phone software app functionality. Our research in hardware-enabled security continued with the development of guidelines on mechanisms that measure and report the security state of BIOS. Our research also continued to expand into the cybersecurity aspects of cyber-physical and embedded systems, and the mechanisms for enabling and protecting public safety communications.

Our ability to interact with the broad federal community continues to be critical to our success. This interaction helps to ensure that our research is consistent with national objectives related to or impacted by information security. This interaction is most prominent in our strengthened collaborations with the Department of Defense, the Intelligence Community, and the Committee on National Security Systems to establish a common foundation for information security across the federal government. The release of Special Publication 800-30, *Guide for Conducting Risk Assessments*, developed by the Joint Task Force Transformation Initiative Interagency Working Group, is not only leading to more consistent ways to assess risks, but is also enabling more effective risk management and facilitating greater sharing of information across organizational boundaries.

For many years, the Computer Security Division (CSD), in collaboration with our global partners from government, industry, and academia, has made great contributions to help secure the nation's critical information and infrastructure. We look forward to furthering these relationships in 2013 as we lead in examining the diverse cybersecurity aspects of a broad set of areas, including supply chain risk management; security analytics; cloud, mobile, and privacy-enhancing technologies; hardware-enabled security; and cyber-physical and embedded systems.

Donna Dodson
Chief, Computer Security Division
& Deputy Chief Cybersecurity Advisor

Donna Dodson

Chief, Computer Security Division, Acting
Associate Director & Acting Chief Cybersecurity
Advisor, Cybersecurity Advisor Office,
and Acting Executive Director, National
Cybersecurity Center of Excellence

Matthew Scholl

Deputy Chief, Computer Security Division
and Acting Associate Director of Operations,
National Cybersecurity Center of Excellence

GROUP MANAGERS

Tim Polk

Cryptographic Technology
Group

Mark (Lee) Badger

Security Components and
Mechanisms Group

David Ferraiolo

Secure Systems and Applications
Group

Kevin Stine

Security Outreach
and Integration Group

Michael Cooper

Security Testing, Validation
and Measurement Group

Cryptographic Technology Group

Mission Statement:

Research, develop, engineer, and standardize cryptographic algorithms, methods, and protocols.

Overview:

The Cryptographic Technology Group's (CTG) work in cryptographic mechanisms addresses topics such as hash algorithms, symmetric and asymmetric cryptographic techniques, key management, authentication, and random number generation. Strong cryptography is used to improve the security of information systems and the information they process. Users then take advantage of the availability of secure applications in the marketplace made possible by the appropriate use of standardized, high-quality cryptography.

CTG continued its work in a number of cryptographic areas, including:

* Specifying and using cryptographic algorithms;
* Revising Federal Information Processing Standard (FIPS) 140-2, *Security Requirements for Cryptographic Modules;*
* Developing hardware roots of trust to support reliable device authentication and establish new bases for system measurement;
* Supporting the Smart Grid Interoperability Panel by assessing the security of cryptographic methods used in the security protocols for the communication and management networks used by public utility companies, and in developing security guidelines for the Smart Grid;
* Advising the FIPS 140-2 Validation Program in the validation of cryptographic algorithms and cryptographic modules; and
* Supporting NIST's Personal Identity Verification (PIV) project that was initiated in response to the Homeland Security Presidential Directive 12 (HSPD-12).

A major highlight of this year is the completion of the SHA-3 Cryptographic Hash Algorithm Competition and the selection of KECCAK as the SHA-3 algorithm.

CTG continued to make an impact in the field of cryptography, both within and outside the federal government, by collaborating with national and international agencies, academic and research organizations, and standards bodies to develop interoperable security standards and guidelines. Federal agency collaborators include the National Security Agency (NSA), the National Telecommunications and Information Administration (NTIA), the General Services Administration (GSA), the Election Assistance Commission (EAC) and the Federal Voting Assistance Program (FVAP). International agencies include the Communications Security Establishment of Canada, and Australia's Defense Signals Agency and Centrelink. National and international standards bodies include the American Standards Committee (ASC) X9 (financial industry standards), the International Organization for Standardization (ISO), the Institute of Electrical and Electronics Engineers (IEEE), the Internet Engineering Task Force (IETF), the American National Standards Institute (ANSI), and the Trusted Computing Group (TCG). Industry collaborators include Intel, Dell, Hewlett Packard, VeriSign, Certicom, Entrust Technologies, Microsoft, Orion Security, RSA Security, Voltage Security, Verifone, Juniper, NTRU Cryptosystems, and Cisco. Academic collaborators include Katholieke Universiteit Leuven (KU Leuven), George Mason University, Danmarks Tekniske Universitet, George Washington University, SDU Odense (University of Southern Denmark), University of California Davis, Malaga University, and Yale University. Academic and research organizations include the International Association for Cryptologic Research (IACR), the European Network of Excellence in Cryptology (ECRYPT) II, and the Japanese Cryptography Research and Evaluation Committees (CRYPTREC).

Group Manager:
Mr. William (Tim) Polk
william.polk@nist.gov
(301) 975-3348

Security Components and Mechanisms Group

Mission Statement:
Research, develop, and standardize foundational security mechanisms, protocols, and services.

Overview:
The Security Components and Mechanisms Group's (SCMG) security research focuses on the development and management of foundational building-block security mechanisms and techniques that can be integrated into a wide variety of mission-critical U.S. information systems. The group's work spans a spectrum from near-term hardening and improvement to the design and analysis of next-generation, leap-ahead security capabilities. Computer security depends fundamentally on the level of trust that can be established for computer software and systems. This work, therefore, focuses strongly on assurance-building activities ranging from the analysis of software configuration settings to advanced trust architectures to testing tools that surface flaws in software modules. Due to the often manual and costly nature of assurance building using current techniques, this work focuses strongly on increasing the applicability and effectiveness of automated techniques wherever feasible. SCMG conducts research collaboratively with government, industry, and academia. The outputs of this research consist of prototype systems, software tools, demonstrations, NIST Special Publications and NIST Interagency Reports, conference papers, and journal papers.

SCMG works on a variety of topics, such as specifications for the automated exchange of security information between systems, computer security incident handling guidelines, formulation of high-assurance software configuration settings, hardware roots of trust for mobile devices, combinatorial testing techniques, conformity assessment of software implementing biometric standards, and adoption of Internet Protocol Version 6 and Internet Protocol security extensions. SCMG collaborates extensively with government, academia, and the private sector. In the last year, collaborations have included the University of Texas Arlington, University of North Texas, University of Maryland-Baltimore County (UMBC), North Carolina State University, Johns Hopkins Applied Physics Lab (APL), National Aeronautics and Space Administration (NASA), U.S. Air Force T&E, Massachusetts Institute of Technology (MIT) Lincoln Labs, the Defense Advanced Research Projects Agency (DARPA), and Carnegie Mellon University (CMU).

Example successes from this work include a second revision of the NIST Special Publication 800-61, *Computer Security Incident Handling Guide*, the Advanced Combinatorial Testing System (ACTS) software and documentation, the NIST BioCTS 2012 conformance testing tool and test assertions, a design for continuous monitoring systems, and a continuous monitoring prototype system.

Group Manager:
Mr. Mark (Lee) Badger
mark.badger@nist.gov
(301) 975-3176

Secure Systems and Applications Group

Mission Statement:
Integrate and apply security technologies, standards, and guidelines for computing platforms and information systems.

Overview:
The Secure Systems and Applications Group's (SSAG) security research focuses on identifying emerging and high-priority technologies, and on developing security solutions that will have a high impact on the U.S. critical information infrastructure. The group conducts research and development on behalf of government and industry from the earliest stages of technology development through proof-of-concept, reference and prototype implementations, and demonstrations. SSAG works to transfer new technologies to industry, produce new standards and guidance for federal agencies and industry, and develop tests, test methodologies, and assurance methods.

Some of the many topics the group investigates include mobile device security, cloud computing and virtualization, identity management, access control and authorization management, and software assurance. This research helps to meet federal information security requirements that may not be fully addressed by existing technology. SSAG collaborates extensively with government, academia, and private sector entities, including FY2012 collaborations with the National Security Agency (NSA), the Department of Defense (DoD), the Defense Advanced Research Projects Agency (DARPA), the Department of Homeland Security (DHS), the White House Communications Agency (WHCA), George Mason University, North Carolina State University, Microsoft Corporation, VM Ware, Symantec, Mobile System 7, One Enterprise Consulting Group, and MITRE.

Example successes from this work include tools for access control policy testing; new concepts in access control and policy enforcement; methods for achieving comprehensive policy enforcement and data interoperability across enterprise data services; test methods for mobile device (smart phone) application security; validation of cryptography for smart phones, and several government wide technical exchange meetings on mobile device security. For the federal government's cloud computing initiatives, SSAG led the NIST Security Working Group's task of developing the *NIST Cloud Computing - Security Reference Architecture* working document, and contributed to the development of the white paper *"Challenging Security Requirements for the USG Cloud Computing Adoption."* To improve access to new technologies, SSAG chaired, edited, and participated in the development of a wide variety of national and international security standards.

Group Manager:
Mr. David Ferraiolo
david.ferraiolo@nist.gov
(301) 975-3046

Security Outreach and Integration Group

Mission Statement:
Develop, integrate, and promote the mission-specific application of information security standards, guidelines, best practices, and technologies.

Overview:
The U.S. economy, citizens, and government rely on information technology. Protecting information technology, including its information and the information infrastructure, is critical for the nation. As part of the Computer Security Division, the Security Outreach and Integration Group (SOIG) leverages its broad cybersecurity and risk management expertise to develop, integrate, and promote security standards, guidelines, tools, technologies, methodologies, tests, and measurements to address critical cybersecurity needs in many areas of national and international importance.

Major initiatives in this area include the development of standards, guidelines, tools, technologies, and methodologies to enable the security and risk management of the Information and Communications Technologies (ICT) supply chain; Smart Grid; Electronic Voting; Cyber Physical and Industrial Control Systems; Health Information Technology; National Public Safety Broadband Network, and the Federal Information Security Management Act (FISMA) implementation program. The group also serves as lead for the National Initiative for Cybersecurity Education (NICE), and provides extended outreach initiatives to stakeholders across federal, state, and local governments, industry, academia, small businesses, and the public.

Key to the group's success is the ability to interact with a broad constituency - government, industry, academia, and the public - in order to ensure that SOIG's program is consistent with national objectives related to or impacted by information security. Through collaboration, cooperation, and open and transparent public engagement, SOIG works to address these critical cybersecurity challenges, enable greater U.S. industrial competitiveness, and facilitate the practical implementation of scalable and sustainable information security standards and practices.

Group Manager:
Mr. Kevin Stine
kevin.stine@nist.gov
(301) 975-4483

Security Testing, Validation, and Measurement Group

Mission Statement:

Advance information security testing, measurement science, and conformance.

Overview:

Federal agencies, industry, and the public rely on cryptography for the protection of information and communications used in electronic commerce, critical infrastructure, and other application areas. At the core of all products offering cryptographic services is the cryptographic module. Cryptographic modules, which contain cryptographic algorithms, are used in products and systems to provide security services such as confidentiality, integrity, and authentication. Although cryptography is used to provide security, weaknesses such as poor design or weak algorithms can render a product insecure and place highly sensitive information at risk. When protecting their sensitive data, federal government agencies require a minimum level of assurance that cryptographic products meet their security requirements. Also, federal agencies are required to use only tested and validated cryptographic modules. Adequate testing and validation of the cryptographic module and its underlying cryptographic algorithms against established standards is essential to provide security assurance.

The group's testing-focused activities include validating cryptographic algorithm implementations, cryptographic modules, and Security Content Automation Protocol (SCAP)-compliant products; developing test suites and test methods; providing implementation guidance and technical support to industry forums; and conducting education, training, and outreach programs.

All of the Security Testing, Validation, and Measurement Group's validation programs work together with independent Cryptographic and Security Testing laboratories that are accredited by the NIST National Voluntary Laboratory Accreditation Program (NVLAP). Based on the independent laboratory test report and test evidence, the Validation Program then validates the implementation under test. The validations awarded to vendor implementations are publicly posted on the NIST website.

Group Manager:
Mr. Michael Cooper
michael.cooper@nist.gov
(301) 975-8077

The E-Government Act, Public Law 107-347, passed by the 107th Congress and signed into law by the President in December 2002, recognized the importance of information security to the economic and national security interests of the United States. Title III of the E-Government Act, entitled the Federal Information Security Management Act (FISMA) of 2002, included duties and responsibilities for the National Institute of Standards and Technology, Information Technology Laboratory, Computer Security Division (CSD). In 2012, CSD addressed its assignments through the following activities:

* Issued one final and two draft Federal Information Processing Standards (FIPS) that specify hash algorithms used to generate message digests, algorithms used to generate digital signatures, and technical requirements for a common identification standard for federal employees and contractors;

* Issued 28 draft and final NIST Special Publications (SPs) that provide management, operational, and technical security guidelines in areas such as BIOS management and measurement, key management and derivation, media sanitization, electronic authentication, security automation, Bluetooth and wireless protocols, incident handling and intrusion detection, malware, cloud computing, public key infrastructure, and risk assessments. In addition, 12 draft and final NIST Interagency Reports were issued on a variety of topics including supply chain risk management, personal identity verification, access control, security automation and continuous monitoring, and the Smart Grid Advanced Metering Infrastructure;

* Produced guidelines concerning the handling of information security incidents to help agencies analyze incident-related data and determine the appropriate response to each incident;

* Continued the successful collaboration with the Office of the Director of National Intelligence, the Committee on National Security Systems, and the Department of Defense to establish a common foundation for information security across the federal government, including a structured, yet flexible approach for managing information security risk across an organization. In 2012, this collaboration produced foundational guidelines for conducting risk assessments, and updated guidelines for selecting and specifying security controls for federal information systems and organizations;

* Provided assistance to agencies and the private sector: conducted ongoing, substantial reimbursable and non-reimbursable assistance to the government and private sector, including many outreach efforts through the Federal Information Systems Security Educators' Association (FISSEA), the Federal Computer Security Program Managers' Forum, and the Small Business Information Security Corner;

* Reviewed security policies and technologies from the private sector and national security systems for potential federal agency use: hosted a repository of federal agency security practices, public/private security practices, and security configuration checklists for IT products. Continued to lead, in conjunction with the Government of Canada's Communications Security Establishment, the Cryptographic Module Validation Program (CMVP). The Common Criteria Evaluation and Validation Scheme (CCEVS) and CMVP facilitate security testing of IT products usable by the federal government;

* Solicited recommendations of the Information Security and Privacy Advisory Board (ISPAB) on draft standards and guidelines, and on information security and privacy issues;

* Conducted workshops, awareness briefings, and outreach to CSD customers to ensure comprehension of standards and guidelines, to share ongoing and planned activities, and to aid in scoping guidelines in a collaborative, open, and transparent manner. CSD also held workshops on diverse information security and technology topics including security automation, identity management, information and communications technologies supply chain risk management, information security awareness and training, cybersecurity of cyber physical systems, technical aspects of botnets, health information security, and mobile computing; and

* Produced an annual report as a NIST Special Publication (SP). The 2003-2011 Annual Reports are available on the Computer Security Resource Center (CSRC) at http://csrc.nist.gov.

CSD Work in National and International Standards

CSD's Part in National and International ISO Security Standards Processes

Figure 1 (below) shows the many national and international standards-developing organizations (SDOs) involved in cybersecurity standardization. CSD participates in many cybersecurity standards activities in many of these organizations, either in leadership positions or as editors and contributors. Many of CSD's publications have been the basis for both national and international standards projects. This section discusses CSD standards activities in conjunction with InterNational Committee for Information Technology Standards (INCITS) Technical Committee Cyber Security (CS) 1, where Dan Benigni serves as Chair and U.S. Head of Delegation to SC 27.

The International Organization for Standardization

The International Organization for Standardization (ISO) is a network of the national standards institutes of 148 countries, with the representation of one member per country. The scope of ISO covers standardization in all fields except electrical and electronic engineering standards, which are the responsibility of the International Electrotechnical Commission (IEC).

The IEC prepares and publishes international standards for all electrical, electronic, and related technologies, including electronics, magnetics and electromagnetics, electroacoustics, multimedia, telecommunication, and energy production and distribution, as well as associated general disciplines such as terminology and symbols, electromagnetic compatibility, measurement and performance, dependability, design and development, safety, and the environment.

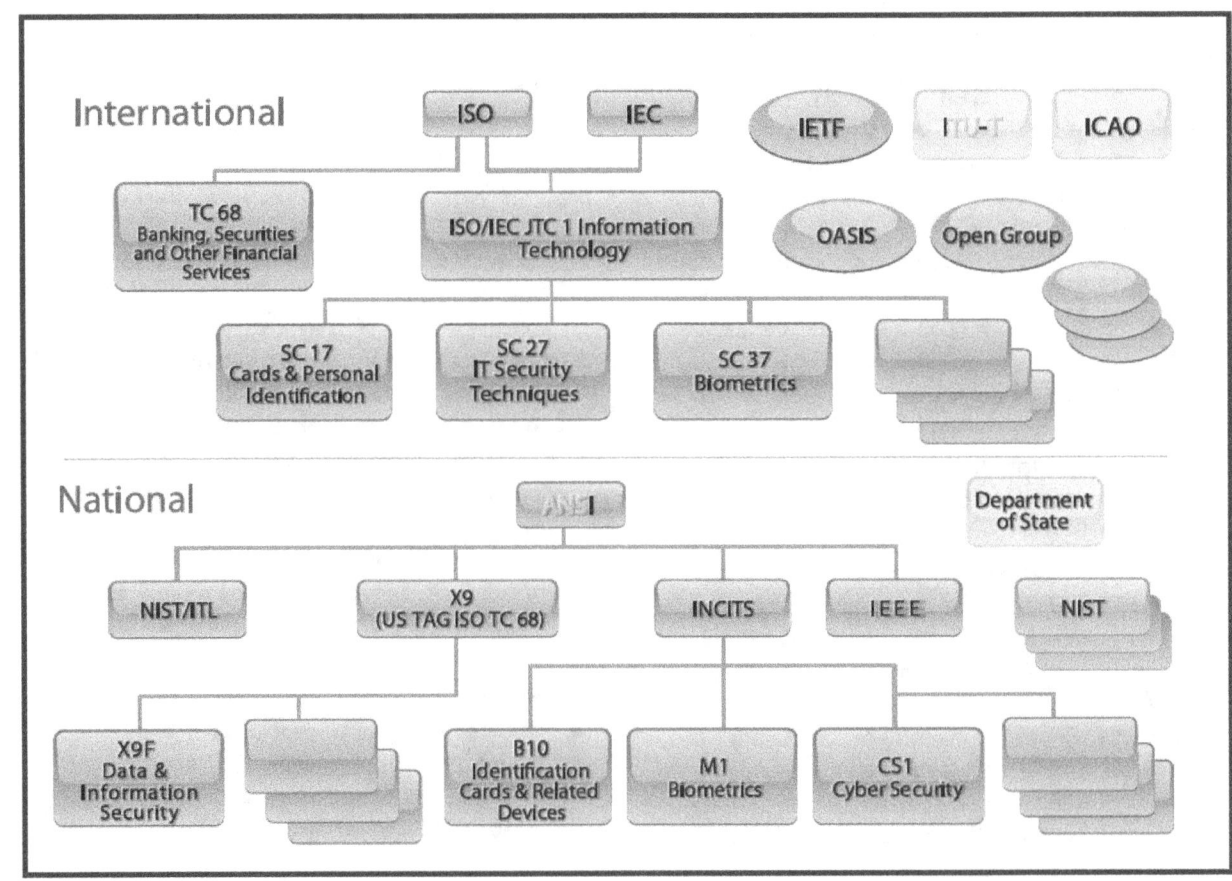

Figure 1: SDOs involved in Cybersecurity

Joint Technical Committee 1 (JTC 1) was formed by ISO and IEC to be responsible for international standardization in the field of Information Technology. It develops, maintains, promotes, and facilitates IT standards required by global markets, meeting business and user requirements concerning—

- Design and development of IT systems and tools;
- Performance and quality of IT products and systems;
- Security of IT systems and information;
- Portability of application programs;
- Interoperability of IT products and systems;
- Unified tools and environments;
- Harmonized IT vocabulary; and
- User-friendly and ergonomically designed user interfaces.

JTC 1 consists of a number of subcommittees (SCs) and working groups that address specific technologies. SCs that produce standards relating to IT security include:

- SC 06 - Telecommunications and Information Exchange Between Systems;
- SC 17 - Cards and Personal Identification;
- SC 27 - IT Security Techniques; and
- SC 37 - Biometrics (Fernando Podio of NIST serves as Chair).

JTC1 also has—

- Technical Committee 68 - Financial Services;
- SC 2 - Operations and Procedures including Security;
- SC 4 - Securities;
- SC 6 - Financial Transaction Cards, Related Media and Operations; and
- SC 7 - Software and Systems Engineering.

The American National Standards Institute

The American National Standards Institute (ANSI) is a private, nonprofit organization (501(c)(3)) that administers and coordinates the U.S. voluntary standardization and conformity assessment system.

ANSI facilitates the development of American National Standards (ANSs) by accrediting the procedures of standards-developing organizations (SDOs). The InterNational Committee for Information Technology Standards (INCITS) is accredited by ANSI.

ANSI promotes the use of U.S. standards internationally, advocates U.S. policy and technical positions in international and regional standards organizations, and encourages the adoption of international standards as national standards where they meet the needs of the user community. ANSI is the sole U.S. representative and dues-paying member of the two major non-treaty international standards organizations, ISO and, via the United States National Committee (USNC), the IEC.

INCITS serves as the ANSI Technical Advisory Group (TAG) for ISO/IEC Joint Technical Committee 1. INCITS is sponsored by the Information Technology Industry (ITI) Council, a trade association representing the leading U.S. providers of information technology products and services. INCITS currently has more than 975 published standards.

INCITS is organized into Technical Committees that focus on the creation of standards for different technology areas. Technical committees that focus on IT security and IT security-related technologies, or that may require separate security standards include:

- B10 - Identification Cards and Related Devices (Sal Francomacaro, Chair Task Group B10.12, Integrated Circuit Cards with Contacts);
- CS1 - Cyber Security (Dan Benigni, Chair and Richard Kissel, NIST Principal voting member);
- E22 - Item Authentication;
- M1 - Biometrics (Fernando Podio, Chair);
- T3 - Open Distributed Processing (ODP);
- T6 - Radio Frequency Identification (RFID) Technology;
- CGIT1 - Corporate Governance of IT (Richard Kissel, NIST Principal voting member and International Representative); and
- DAPS38 - Distributed Application Platforms and Services.

As a technical committee of INCITS, CS1 develops United States, national, ANSI-accredited standards in the area of cybersecurity. Its scope encompasses—

* Management of information security and systems;
* Management of third-party information security service providers;
* Intrusion detection;
* Network security;
* Cloud computing security;
* Supply chain risk management;
* Incident handling;
* IT security evaluation and assurance;
* Security assessment of operational systems;
* Security requirements for cryptographic modules;
* Protection profiles;
* Role-based access control;
* Security checklists;
* Security metrics;
* Cryptographic and non-cryptographic techniques and mechanisms including confidentiality, entity authentication, non-repudiation, key management, data integrity, message authentication, hash functions, and digital signatures;
* Future service and applications standards supporting the implementation of control objectives and controls as defined in ISO 27001, in the areas of business continuity, and outsourcing;
* Identity management, including identity management framework, role-based access control, and single sign-on; and
* Privacy technologies, including privacy framework, privacy reference architecture, privacy infrastructure, anonymity and credentials, and specific privacy-enhancing technologies.

The scope of CS1 explicitly excludes the areas of work on cybersecurity standardization presently under way in INCITS B10, M1, T3, T10, and T11, as well as other standard groups, such as the Alliance for Telecommunications Industry Solutions (ATIS), the

Institute of Electrical and Electronics Engineers, Inc. (IEEE), the Internet Engineering Task Force (IETF), the Travel Industry Association of America (TIAA), and Accredited Standards Committee (ASC) X9. The CS1 scope of work includes standardization in most of the same cybersecurity areas as are covered in the NIST CSD.

As the U.S. TAG to ISO/IEC JTC 1/SC 27, CS1 contributes to the SC 27 program of work on IT Security Techniques in terms of comments and contributions on SC 27 standards projects; votes on SC 27 standards documents at various stages of development; and nominates U.S. experts to work on various SC 27 projects as editors, coeditors, or in other SC 27 leadership positions. Currently a number of CS1 members are serving as SC 27 document editors or coeditors on various standards projects, including CSD staff Randy Easter and Richard Kissel.

All input from CS1 is processed through INCITS to ANSI, then to SC 27. It is also a conduit for getting U.S. - based new work item proposals and U.S.-developed national standards into the international SC 27 standards development process. In its international efforts, CS1 responded to all calls for U.S. contributions and/or voting positions on all international security standards projects in ISO/IEC JTC1 SC 27 in a consistent, efficient, and timely manner.

NIST contributes to many of CS1's national and international IT security standards efforts through its membership on CS1, where Dan Benigni serves as the nonvoting chair and Richard Kissel as the NIST Principal voting member. Internationally, there are over 100 published standards, and almost all have been adopted as U.S. national standards. There are more than 100 current international standards projects.

CSD's Role in Cybersecurity Standardization

CSD's cybersecurity research also plays a direct role in the Cybersecurity Standardization efforts of CS1 at the national level. During FY2012:

* Through the CS1 Ad hoc Group on Role-Based Access Control (RBAC), ANSI has published three national standards:

- "Requirements for the Implementation and Interoperability of Role Based Access Control";
- "Revision of INCITS 359 - 2004, Role Based Access Control (RBAC)"; and
- "Information technology -- Role Based Access Control - Policy Enhanced" and Project 2214-D.

NIST originally authored RBAC, and Rick Kuhn is the NIST lead in the Ad hoc group.

* The NIST Policy Machine research and development has resulted in three ongoing national standards projects in CS1 in the early stages of development. They include:
  - "Next Generation Access Control -Implementation Requirements, Protocols and API Definitions (NGAC-IRPADS)." Its assigned project number is 2193-D;
  - Just published "Next Generation Access Control -Functional Architecture (NGAC-FA)." Its assigned project number is 2194-D; David Ferraiolo of NIST is editor; and
  - "Next Generation Access Control - Generic Operations & Abstract Data Structures (NGAC-GOADS)." Its assigned project number is 2195-D, and Serban Gavrila of NIST is editor.

Within CS1, liaisons are maintained with nearly 20 organizations. They include the following:

* Open Group;
* IEEE P1700 and P1619;
* Forum of Incident Response and Security Teams (FIRST);
* American Bar Association (ABA), section on Science and Technology;
* ABA Federated Identity Management Legal (IdM Legal) Task Force;
* INCITS T11, M1, CGIT1, DAPS38, and PL22;
* Internet Security Alliance;
* Trusted Computing Group;
* Kantara Initiative Identity Assurance Working Group (IAWG);
* Cloud Security Alliance;

* SC 7 TAG;
* Scientific Working Group on Digital Evidence (SWGDE); and
* The Storage Networking Industry Association (SNIA).

Dan Benigni also serves as cybersecurity standards coordinator in CSD.

Contact:
Mr. Daniel Benigni
(301) 975-3279
benigni@nist.gov

Identity Management Standards within INCITS B10 and ISO JTC1/SC 17

CSD supports Identity Management standardization activities through participation in national and international standards bodies and organizations. CSD actively participates in InterNational Committee for Information Technology Standards (INCITS) B10 committee which is focused on interoperability of Identification Cards and Related Devices. CSD staff serves as Chair and Vice Chair of the B10.12 committee which develops interoperable standards for Integrated Circuit Cards with Contacts. CSD staff also serves as the U.S. Head of delegation to Working Group 4 and Working Group 11 of the International Organization for Standardization / International Electrotechnical Commission (ISO/IEC) Joint Technical Committee (JTC) 1 Subcommittee 17.

In addition to chairing the B10.12 committee, CSD provides technical and editorial support in the development of national and international standards. Specifically, CSD staff serves as the technical editor of ANSI 504-1, Generic Identity Command Set (GICS). GICS provides for Personal Identity Verification (PIV), PIV-I (PIV-Interoperable) and Common Access Card (CAC) card applications (but not limited to these applications) to be built from a single platform. GICS defines an open platform where identity applications can be instantiated, deployed, and used in an interoperable way between the credential issuers and credential users. CSD staff also provides significant input to standards of major interest to U.S. government agencies and U.S. markets. CSD influences the development and revision of ISO/IEC 7816 (Identification Cards, Integrated Circuit Cards), ISO/IEC 24727 (Identification Cards, Integrated Circuit Card Programming Interfaces), and ISO/IEC 24787 (Biometrics 'Match On Card' Comparison).

During FY2012, many of these standards reached important milestones: ANSI 504 was published; ISO/IEC 24727 amendments to part 2, 3, and 4 were published; ISO/IEC 7816 part 4 was updated to add significant functionality: several modifications requested by the U.S. delegation are now part of this standard. CSD provides contributions and feedback on many other INCITS B10 standards projects that support identity management.

CSD's investment in this section is motivated by the "new technical ideas" that are supported by these standards. For example, ANSI 504 (GICS) is an ID platform that leverages the existing FIPS 201 (PIV) infrastructure to offer support to a larger number of initiatives for both government agencies and enterprises. In particular, it aims to support initiatives like the National Strategy for Trusted Identities in Cyberspace (NSTIC). ISO/IEC 24727 aims to create an interoperability framework to increase resilience and scalability of identity management solutions and provide interoperability domestically and internationally.

Contact:

Mr. Salvatore Francomacaro

(301) 975-6414

salvatore.francomacaro@nist.gov

Federal Information Security Management Act (FISMA) Implementation Project

The FISMA Implementation Project focuses on:

* Developing a comprehensive series of standards and guidelines to help federal agencies build strong cybersecurity programs, defend against increasingly sophisticated cyber attacks, and demonstrate compliance to security requirements set forth in legislation, Executive Orders, Homeland Security Directives, and Office of Management and Budget (OMB) polices;

* Building common understanding and reference guides for organizations applying the NIST suite of standards and guidelines that support the NIST Risk Management Framework (RMF);

* Developing minimum criteria and guidelines for recognizing security assessment organization provider's as capable of assessing information systems consistent with NIST standards and guidelines supporting the RMF; and

* Conducting FISMA outreach to public and private sector organizations.

During 2011-2012, CSD strengthened its collaboration with the Department of Defense, the Intelligence Community, and the Committee on National Security Systems, in partnership with the Joint Task Force Transformation Initiative, which continues to develop key cybersecurity guidelines for protecting federal information and information systems for the Unified Information Security Framework. Previously, the Joint Task Force developed common security guidance in the critical areas of security controls for information systems and organizations, security assessment procedures to demonstrate security control effectiveness, security authorizations for risk acceptance decisions, and continuous monitoring activities to ensure that decision makers receive the most up-to-date information on the security state of their information systems. In addition, CSD worked with the General Services Administration's (GSAs), Federal Risk and Authorization Management Program to identify security assessment requirements, and prototype a process for approving Third-Party Assessment Organizations (3PAOs) that demonstrate capability in assessing Cloud Service Providers (CSPs) information systems for conformance to NIST standards and guidelines.

In FY2012, CSD worked on the following three initiatives:

(i) *Risk Management and Risk Assessment Guidelines:* Developed a comprehensive risk assessment guideline examining the relationships among key risk factors including threats, vulnerabilities, impact, and likelihood. NIST Special Publication (SP) 800-30, Revision 1, *Guide for Conducting Risk Assessments*, is the fifth in the series of risk management and information security guidelines being developed by the Joint Task Force. This revision changes the focus of SP 800-30, originally published as a risk management guideline, to focus exclusively on conducting risk assessments. The risk assessment guidance in SP 800-30 has been significantly expanded to include more in-depth information

on a wide variety of risk factors essential to determining information security risk (e.g., threat sources and events, vulnerabilities and predisposing conditions, impact, and likelihood of threat occurrence). A three-step process is described including key activities to prepare for risk assessments, activities to successfully conduct risk assessments, and approaches to maintain the currency of assessment results.

In addition to providing a comprehensive process for assessing information security risk, this publication also describes how to apply the process at the three tiers in the risk management hierarchy—the organization level, the mission/business process level, and the information system level. To facilitate ease of use for individuals or groups conducting risk assessments within organizations, this publication also provides a set of exemplary templates, tables, and assessment scales for common risk factors. The templates, tables, and assessment scales give maximum flexibility in designing risk assessments based on the express

GENERIC RISK MODEL WITH KEY RISK FACTORS

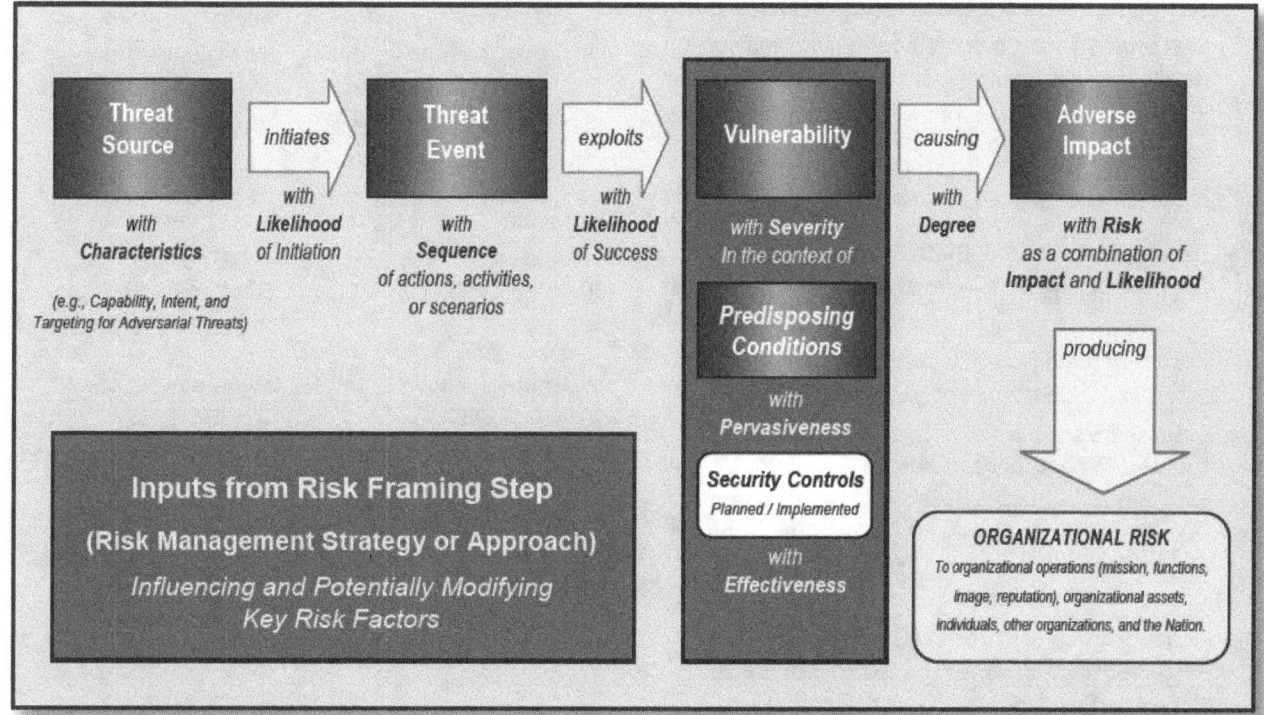

Figure 2

purpose, scope, assumptions, and constraints established by organizations. The Figure 2 on previous page illustrates a generic risk model with key risk factors.

(ii) *Criteria and Guidelines for Recognizing Security Assessment Provider Organizations:* CSD developed proficiency tests and capability demonstration requirements for evaluating the capability of security assessment organizations to conduct security assessments of information systems security controls for compliance to NIST standards and guidelines. The technical capability requirements were derived from draft NIST Interagency Report (NISTIR) 7328, *Security Assessment Provider Requirements and Customer Responsibilities: Building a Security Assessment Credentialing Program for Federal Information Systems.*

(iii) *FISMA Outreach Activity to Public and Private Sector Organizations:* CSD conducted cybersecurity outreach briefings and provided support to state and local governments as well as private sector organizations on topics of interest, such as effective implementation of the NIST Risk Management Framework. In addition, CSD conducted outreach activities with academic institutions, providing information on NIST's security standards and guidelines and exploring new areas of cybersecurity research and development.

In FY2012, CSD completed the following activities:

* Developed NIST SP 800-30, Revision 1, *Guide for Conducting Risk Assessments;*

* Developed draft SP 800-53, Revision 4, *Recommended Security Controls for Federal Information Systems and Organizations;*

* Released example security assessment cases for assessment procedures in SP 800-53A, Revision 1, *Guide for Security Assessment of Federal Information Systems and Organizations: Building Effective Security Assessment Plans;* and

* Collaborated with the ITL Software and Systems Division and the NIST Standards Coordination Office using the International

Standard ISO/IEC 17020:2008 *General criteria for the operation of various types of bodies performing inspections,* in support of GSA in establishing a process for qualifying Third Party Assessment Organizations (3PAOs) to conduct security assessments of Cloud Service Providers (CSPs) information systems consistent with GSA requirements based on NIST standards and guidelines.

In FY2013, CSD intends to:

* Finalize SP 800-53, Revision 4, *Recommended Security Controls for Federal Information Systems and Organizations;*

* Develop SP 800-53A, Revision 2, *Guide for Assessing the Security and Privacy Controls in Federal Information Systems and Organizations;*

* Develop draft SP 800-18, Revision 2, *Guide for Developing Security Plans for Federal Information Systems and Organizations;*

* Develop an information system security and engineering guideline; and

* Expand cybersecurity outreach to include additional state, local, and tribal governments as well as private sector organizations and academic institutions. Additionally, the outreach program will continue to support federal agencies in effective implementation of the NIST Risk Management Framework.

http://csrc.nist.gov/sec-cert

Contacts:

Dr. Ron Ross
(301) 975-5390
ron.ross@nist.gov

Ms. Pat Toth
(301) 975-5140
patricia.toth@nist.gov

Mr. Arnold Johnson
(301) 975-3247
arnold.johnson@nist.gov

Ms. Kelley Dempsey
(301) 975-2827
kelley.dempsey@nist.gov

Ms. Peggy Himes
(301) 975-2489
peggy.himes@nist.gov

Biometric Standards and Associated Conformity Assessment Testing Tools

CSD staff responds to government, industry, and market requirements for open systems standards by:

* Accelerating development of formal biometric standards;
* Providing effective leadership and technical participation in the development of these standards;
* Developing Conformance Test Architectures and Test Suites designed to test implementations of biometric standards;
* Supporting harmonization of biometric, tokens, and security standards;
* Promoting biometric standards adoption; and
* Promoting conformity assessment efforts.

CSD continues to work in close partnership with government agencies, industry, and academic institutions to achieve the project goals delineated above. CSD actively participates in a number of biometric standards development projects, contributes to the development of biometric standards, and leads national and international biometric standards bodies. Nationally, CSD's staff leads InterNational Committee for Information Technology Standards (INCITS) Technical Committee 1 (M1) - *Biometrics*; international efforts include Joint Technical Committee 1 (JTC 1) of the International Standards Organization (ISO) and the International Electrotechnical Commission (IEC) Subcommittee SC 37 - Biometrics - JTC 1/SC 37. CSD plans to continue this work in FY2013.

During FY2012, CSD developed and publicly released one advanced Conformance Test Architecture (CTA) designed to test implementations of the International Standards Organization/International Electrotechnical Commission (ISO/IEC) biometric standards called *BioCTS2012 for ISO/IEC* and another CTA designed to test implementations of the American National Standards Institute (ANSI)/NIST-ITL 1-2011 (AN-2011) standard.[1] This architecture is called *BioCTS2012 for ANSI/NIST 2011*.

Eight Conformance Test Suites (CTSs) to test implementations of binary-encoded data formats were developed and released. These test implementations of the first and second generation of binary-encoded data formats developed by the Joint Technical Commission 1 Subcommittee 37 - *Biometrics* (JTC 1/SC 37) (finger minutiae and finger image, face, and iris image data formats). The CTS set also includes two CTSs designed to support testing of on-card and off-card Personal Identity Verification (PIV) iris data format profiles specified in draft NIST Special Publication (SP) 800-76-2, *Biometric Data Specification for Personal Identity Verification* (released June 2012). These CTSs run under BioCTS 2012 for ISO/IEC.

BioCTS2012 for AN-2011 incorporates over 1,200 test assertions designed to test the requirements of selected Record Types of AN-2011. Other Record Types included in AN-2011 Transactions are detected and their length and location in the transactions are reported. The test assertions implemented in code for BioCTS2012 for AN-2011 are documented in NIST SP 500-295, *Conformance Testing Methodology for ANSI/NIST-ITL 2011, American National Standard for Information Systems, Data Format for the Interchange of Fingerprint, Facial & Other Biometric Information* (September 2012), developed by the NIST/ITL AN-2011 Conformance Testing Methodology Working Group.

These tools were designed to achieve significant functionality, usability, and performance. NIST Interagency Report (NISTIR) 7877, BioCTS 2012: *Advanced Conformance Test Architectures and Test Suites for Biometric Data Interchange Formats and Biometric Information Records* (September 2012), and a presentation delivered at the last Biometric Consortium conference[2] discussed these test tools and provided technical implementation details. This work is sponsored, in part, by DHS/US-VISIT.

BioCTS 2012

Biometric Conformance Test Software

Iris, Face, Finger ...   Standard Data Interchange Formats   Results
(ISO/IEC, ANSI/NIST, BIRs ...)

Does YOUR Data Conform? Find Out.

In FY2013, CSD plans to develop additional CTSs to test implementations of international biometric data interchange formats under development in JTC 1/SC 37 for both binary and Extensible Markup Language (XML) data-encoded formats as well as extending BioCTS for AN-2011 to other record types (for Traditional and XML encoding). CSD plans to research handling large number of files concurrently to increase still further test report capabilities and analysis. CSD will also continue researching and developing additional test environments support (such as web services and tools in the cloud).

A more detailed discussion on the need for conformance testing and roadmaps of conformance test tools development for the following period can be found in NISTIR 7877 and two related presentations delivered during 2012[3, 4.] CSD is currently evaluating the users' needs for conformance test tools to support testing of additional biometric standard profile implementations.

CSD cochaired the Biometric Consortium conferences and developed the conference technical programs in cooperation with a member of the National Security Agency's staff. The 2012 annual conference, held September 18-20, 2012 in Tampa, Florida, included nearly 1,700 attendees from 30 countries representing government, industry, and academia. The four-track conference included presentations given by internationally recognized experts in biometric technologies, system and application developers, IT business strategists, and government and commercial officers. It focused on biometric technologies for homeland security, identity management, border crossing, electronic commerce, and other applications.

Dr. Charles H. Romine, Director of the Information Technology Laboratory at NIST, was a keynote speaker. Other keynote and featured speakers were Mr. Steven Martinez, Executive Assistant Director, Science and Technology Branch, Federal Bureau of Investigation; Mr. Peter F. Verga, Chief of Staff for the Under

Secretary of Defense for Policy; Dr. Husni Fahmi, Chief of Sub-directorate Population Administration, Head of the e-KPT Technical Team, Ministry of Home Affairs, Indonesia; Mr. Stephen Dennis, Technical Director HSARPA (Homeland Security Advanced Research Projects Agency), Science and Technology Directorate, Department of Homeland Security; Mr. Duane Blackburn, Multi-Discipline Systems Engineer, MITRE; and Mr. R. S. Sharma, Director General and Mission Director, Unique Identification Authority of India.

The conference program included sessions on federal government programs, advances on biometric technologies (e.g., face recognition, rapid DNA [Deoxyribonucleic acid], iris) and standards, Institute of Electrical and Electronics Engineers (IEEE) Biometrics - Identity and Security (BIdS) research, international biometric programs, and an Armed Forces Communications and Electronics Association (AFCEA) Identity Management session. NIST's session highlighted achievements and ongoing biometric research, testing, and standards projects delivered by experts from the NIST Information Access and Computer Security Divisions, the Standards Coordination Office, and the Office of Law Enforcement Standards. In addition, a number of workshops and special sessions were held addressing topics such as biometric security, large-scale identity solutions, Organization for the Advancement of Structured Information Standards (OASIS) biometrics, the role of biometrics in the National Strategy for Trusted Identities in Cyberspace (NSTIC) identity ecosystem, a new biometrics domain for XML, and biometrics in a cloud computing environment. Over 135 speakers participated in the program. CSD coordinated NIST's booth at the Technology Expo presented by AFCEA, which was collocated with the conference. The NIST booth showcased NIST/ITL ongoing projects and provided attendees with technology demonstrations and technical information on these projects. CSD plans to continue supporting the Biometric Consortium at its 2013 annual conference scheduled for September 17-19 in Tampa, Florida.

ITL's Biometric Resource Center:
http://www.nist.gov/biometrics

BioCTS 2012 - Biometric Conformance Test Tool Downloads:
http://www.nist.gov/itl/csd/biometrics/biocta_download.cfm#CTAdownloads

Biometric Consortium website:
http://www.biometrics.org

Biometric Consortium 2012 conference program (released presentations are linked):
http://www.biometrics.org/bc2012/program.pdf

References:

(1) ANSI/NIST-ITL 1-2011, NIST Special Publication 500-290, *Data Format for the Interchange of Fingerprint, Facial & Other Biometric Information*, NIST/ITL, B. Wing (Editor), November 2011.

(2) Conformance Test Architectures for Biometric Data Interchange Formats - Implementation Characteristics, Standards Session, Biometric Consortium conference, D. Yaga and F. Podio, September 2012.

(3) Performance Without Conformance? Value of Level 1, 2 and 3 Conformance, International Biometric Performance Conference, NIST, F. Podio, March 2012: http://biometrics.nist.gov/cs_links/ibpc2012/presentations/Day3/339_podio_IBPC.pdf

(4) Biometric Conformance Test Tools Development Roadmaps, NIST Session, Biometric Consortium conference, F. Podio, September 2012: http://biometrics.org/bc2012/presentations/NIST/PODIOFINALSEPT18NISTSESSION440-455.pdf.

Contact:
Mr. Fernando Podio
(301) 975-2947
fernando.podio@nist.gov

Cybersecurity of Cyber Physical Systems

Since 2009, NIST has been very active in the area of the Smart Grid. The Computer Security Division (CSD) has provided leadership and expertise to the Smart Grid Interoperability Panel's Cyber Security Working Group. Leveraging CSD's broad expertise in relevant areas, CSD is now looking at the cybersecurity needs of the broader landscape of cyber-physical systems.

Cyber-physical systems (CPS), hybrid networked cyber and engineered physical elements codesigned to create adaptive and predictive systems for enhanced performance, are commonly used in the nation's critical infrastructure. Such systems control the electrical grid, provide clean water, produce chemicals, and underlie transportation systems. CPS are gaining in capability as advances are made in technology, and they are critical to future engines of growth such as advanced manufacturing, as well as safety initiatives such as autonomous driving cars.

Cybersecurity is an important cross-cutting discipline that is critical to provide confidence that CPS and their information and supporting communications and information infrastructure are adequately safeguarded. CPS have many unique characteristics, including the need for real-time response and extremely high availability, predictability, and reliability. However, despite the ubiquity and criticality of CPS, little thought has been given so far to secure design. As a result, there have been successful and major attacks such as Stuxnet, Duqu, Flame, and Gauss that target CPS controlling critical infrastructure.

In April 2012, CSD hosted a two-day workshop to explore CPS cybersecurity needs, with a focus on research results and real-world deployment experiences across multiple industries, including healthcare, manufacturing, automotive, and the electric Smart Grid. The goals of the workshop were to gain a better understanding of the cybersecurity challenges faced by CPS across multiple industries, and to determine if there are security requirements that are unique to CPS as opposed to strictly cyber or physical systems. More information on this workshop, including the agenda, session abstracts, and slides, is available at http://www.nist.gov/itl/csd/cyberphysical-workshop.cfm.

In 2013, CSD plans to host a workshop to identify and evaluate the current and needed CPS standards and practices, as well as tools, technologies, and safeguards that protect CPS. CSD, in conjunction with NIST's Engineering Laboratory, will also be updating SP 800-82, *Guide to Industrial Control Systems (ICS) Security*. CSD will continue participating with the International Society of Automation (ISA) 99 Committee, which develops and establishes standards, recommended practices, technical reports, and related information that define procedures for implementing electronically secure industrial automation and control systems and security practices, and for assessing electronic security performance.

Contacts:

Ms. Tanya Brewer
(301) 975-4534
tbrewer@nist.gov

Ms. Suzanne Lightman
(301) 975-6442
suzanne.lightman@nist.gov

Cybersecurity Research & Development (R&D)

The Networking and Information Technology Research and Development (NITRD) Program provides a framework in which many federal agencies come together to coordinate their networking and information technology (IT) research and development (R&D) efforts. CSD remained committed to the value of communicating its R&D efforts to other federal colleagues and identifying the opportunities to support the R&D efforts of federal colleagues throughout the federal government.

The NITRD Program operates under the aegis of the NITRD Subcommittee of the National Science and Technology Council (NSTC) Committee on Technology. The Subcommittee, made up of representatives from each of NITRD's member agencies, provides overall coordination for NITRD activities.

Federal IT R&D, which launched and fueled the digital revolution, continues to drive innovation in scientific research, national security, communication, and commerce to sustain U.S. technological leadership. The NITRD agencies' collaborative efforts increase the overall effectiveness and productivity of these federal R&D investments, leveraging strengths, avoiding duplication, and increasing interoperability of R&D products.

The NITRD Program has its focus on the following research areas:

* Big Data (BD);
* Cyber-Physical Systems (CPS);
* Cyber Security and Information Assurance (CSIA);
* Health Information Technology Research and Development (Health IT R&D);
* Human Computer Interaction and Information Management (HCI&IM);
* High-Confidence Software and Systems (HCSS);
* High-End Computing (HEC);
* Large-Scale Networking (LSN);
* Software Design and Productivity (SDP);
* Social, Economic, and Workforce Implications of IT and IT Workforce Development (SEW); and
* Wireless Spectrum Research and Development (WSRD).

CSD maintains a strong presence in many of these groups with leadership roles in the CSIA Interagency Working Group (IWG) cochaired by Bill Newhouse and the SEW Education Team cochaired by Dr. Ernest McDuffie. Colleagues from other divisions within the Information Technology Laboratory maintain leadership roles in Faster Administration of Science and Technology Education and Research (FASTER) Community of Practice (CoP), HCI&IM, and SDP. NIST colleagues from the Engineering Laboratory have leadership roles in HCSS and the Senior Steering Group for Cyber-Physical Systems.

The CSIA IWG used its monthly meetings to explore the themes and thrusts expressed in the Strategic Plan for the Federal Cybersecurity Research and Development.[1] This plan represents the culmination of focused interactions with academia and industry on forward-thinking concepts and was published by the CSIA IWG in December 2011. The plan introduces

[1] http://www.whitehouse.gov/sites/default/files/microsites/ostp/fed_cybersecurity_rd_strategic_plan_2011.pdf

four game-changing themes (Tailored Trustworthy Spaces, Moving Target, Cyber Economic Incentives, and Designed-In Security); describes the need to develop Scientific Foundations which minimize future cybersecurity problems by developing the science of security; identifies the relevant national priorities where CSD can maximize research impact by catalyzing coordination, collaboration, and integration of research activities across federal agencies for maximum effectiveness; and accelerates Transition to Practice by expediting improvements in cyberspace from research findings through focused transition programs.

The breadth of CSD's work was shared over the course of the year's monthly meetings. Specific briefings were given on CSD's efforts in the area of Continuous Monitoring, Smart Grid, and Healthcare IT Security.

In addition to coordination via the NITRD programs, CSD is a regular participant in the coordination activities of the Federal Special Cyber Operations Research and Engineering (SCORE) Committee, which was established based on recognition by the architects of the Comprehensive National Cybersecurity Initiative (CNCI) that expanded R&D coordination was vital to allow the nation to leap ahead of today's cybersecurity challenges.

On behalf of SCORE, CSD hosted an October 2011 cybersecurity "assumption buster" workshop focused on cloud computing. It was the fourth in a series of workshops designed to explore the assumptions that cyber space is an adversarial domain, and the adversary is tenacious, clever, and capable. By reexamining cybersecurity solutions in the context of these assumptions, the workshops aimed to identify key insights that will lead to novel solutions for some of the nation's needs.

The SCORE committee interacts with CNCI leadership. CSD not only represents the breadth of its work to this committee, but also shares the progress and relevance of research initiatives from other divisions across NIST's Information Technology Laboratory (ITL), such as *Foundations of Measurement Science for Information System* (Applied and Computational Mathematics Division) and *Usability* (Information Access Division).

ITL has been charged to lead the nation in utilizing existing and emerging IT to meet national priorities that reflect the country's broad-based social, economic, and political values and goals. ITL seeks to scale new frontiers in Information Measurement Science to enable international social, economic, and political advancement by collaborating and partnering with industry, academia, and other NIST laboratories to advance science and engineering, setting standards and requirements for unique scientific instrumentation and experiments, data, and communications. The NIST investment in providing leadership in R&D collaboration, particularly CSD's focus on cybersecurity, remains a focus.

Contacts:

Mr. Bill Newhouse
CSIA IWG, CSIA SSG,
and SCORE rep.
(301) 975-2869
william.newhouse@nist.gov

Dr. Ernest McDuffie
SEW Education Team

(301) 975-8897
ernest.mcduffie@nist.gov

Security Aspects of Electronic Voting

In 2002, Congress passed the Help America Vote Act (HAVA) to encourage the upgrade of voting equipment across the United States. HAVA established the Election Assistance Commission (EAC) and the Technical Guidelines Development Committee (TGDC), chaired by the Director of NIST. HAVA calls on NIST to provide technical support to the EAC and TGDC in efforts related to human factors, security, and laboratory accreditation. As part of NIST's efforts, CSD supports the activities of the EAC and the TGDC related to voting equipment security.

In the past year, CSD supported the EAC in updating the Voluntary Voting System Guidelines (VVSG), VVSG 1.1, by assisting the EAC with the development of a new draft of the guidelines for public comment. The security guidelines were updated to improve the auditability of voting systems, to provide greater software integrity protections, to expand and improve access control requirements, and to help ensure that cryptographic security mechanisms are implemented properly. In addition, CSD supported the efforts of the

EAC and the Federal Voting Assistance Program (FVAP) of the Department of Defense (DoD) to improve the voting process for citizens under the Uniformed and Overseas Citizens Voting Act (UOCAVA) by leveraging electronic technologies. CSD worked with the TDCG's UOCAVA Working Group to develop a narrative risk analysis on current UOCAVA voting processes, including vote-by-mail and electronic ballot delivery. CSD's work on voting technologies has also spun off interesting research topics, including the Rabin Beacon project that is discussed separately in this annual report.

In FY2013, CSD will assist the EAC in developing responses to public comments and provide updates to VVSG 1.1 and the associated security test suites. CSD will continue to support the efforts for the EAC and FVAP to improve the voting process for UOCAVA voters. CSD will also continue to expand on research activities, particularly in the areas of risks to voting systems and innovative voting system architectures.

In addition, CSD will support the NIST National Voluntary Laboratory Accreditation Program (NVLAP) efforts to accredit voting system test laboratories by developing proficiency tests and testing artifacts, and by updating Handbook 150-22 used to accredit voting system test laboratories. CSD plans to engage voting system manufacturers, voting system test laboratories, state election officials, and the academic community to explore ways to increase voting system security and transparency.

http://vote.nist.gov/

Contacts:

Dr. Nelson Hastings
(301) 975-5237
nelson.hastings@nist.gov

Mr. Andrew Regenscheid
(301) 975-5155
andrew.regenscheid@nist.gov

Mr. Joshua Franklin
(301) 975-8463
joshua.franklin@nist.gov

Health Information Technology Security

Health information technology (HIT) makes it possible for healthcare providers to better manage patient care through secure use and sharing of health information, leading to improvements in healthcare quality, reduced medical errors, increased efficiencies in care delivery and administration, and improved population health. Central to reaching these goals is the assurance of the confidentiality, integrity, and availability of health information. CSD works actively with government, industry, academia, and others to provide security tools, technologies, and methodologies that provide for the security and privacy of health information.

In FY2012, NIST issued the Health Insurance Portability and Accountability Act (HIPAA) Security Rule self-assessment toolkit to help organizations better understand the requirements of the HIPAA Security Rule, implement those requirements, and assess those implementations in their operational environment. This project enabled NIST to leverage security automation specifications within the context of the healthcare use case.

NIST continued its HIT security outreach efforts. NIST and the Department of Health and Human Services' (DHHS) Office for Civil Rights (OCR) cohosted the fifth annual HIPAA Security Rule conference, "*Safeguarding Health Information: Building Assurance through HIPAA Security,*" in June 2012 at the Ronald Reagan Building and International Trade Center in Washington, D.C. The conference offered important sessions that focused on broad topics of interest to the healthcare and health IT security community. Over 500 attendees from federal, state, and local governments, academia, HIPAA-covered entities and business associates, industry groups, and vendors heard from, and interacted with, healthcare, security, and privacy experts on technologies and methodologies for safeguarding health information and for implementing the requirements of the HIPAA Security Rule. Presentations covered a variety of current topics including updates on OCR's health information privacy and security audit and enforcement activities; establishing an access audit program; securing mobile devices; securing health information in the Cloud;

improving the usability and accessibility of HIT; managing breaches of health information; and the relationship between the HIPAA Security Rule and Meaningful Use.

In FY2013, NIST plans to issue a draft revision to Special Publication (SP) 800-66, *An Introductory Resource Guide for Implementing the HIPAA Security Rule*. As part of its continued outreach efforts, NIST also plans to host the sixth annual "*Safeguarding Health Information*" conference with OCR.

http://www.nist.gov/healthcare/security/index.cfm

Contacts:

Mr. Kevin Stine	Mr. Matthew Scholl
(301) 975-4483	(301) 975-2941
kevin.stine@nist.gov	matthew.scholl@nist.gov

ICT Supply Chain Risk Management

Federal agency information systems are increasingly at risk of both intentional and unintentional supply chain compromise due to the growing sophistication of information and communications technologies (ICT) and the growing speed and scale of a complex, distributed global supply chain. A lack of visibility into, and control over, the ICT supply chain increases the risk technologies will be vulnerable to a variety of threats, including counterfeit materials, malicious software, disruption in logistics, and makes it increasingly difficult for federal agencies to understand their exposure and manage the associated supply chain risks.

The ICT Supply Chain Risk Management (SCRM) project seeks to provide federal agencies with a standardized, repeatable, and feasible toolkit of technical and intelligence resources to strategically manage supply chain risk throughout the entire life cycle of ICT products, systems and services.

In 2012, CSD continued to collaborate with several industry, standards, academic, and government organizations in order to learn about their efforts related to SCRM, and provide technical input as appropriate. NIST's CSD issued draft NISTIR 7622, *Notional Supply Chain Risk Management Practices for Federal Information Systems*, in FY2012. This draft document provides an initial set of means and methods for reducing the risk associated with ICT supply chain vulnerabilities. It also seeks to equip federal departments and agencies with implementation guidance as well as offer a set of practices acquiring organizations can implement to increase their understanding of, and visibility throughout, the supply chain.

With a grant issued by NIST, the University of Maryland (UMD), Robert H. Smith School of Business, Supply Chain Management Center published a report which inventories the proliferating array of existing public and private sector ICT supply chain initiatives across diverse ICT segment and formulates a framework for defining various initiative within a single SCRM architecture. This framework has three tiers: *enterprise risk governance, system integration*, and *operations*. Within each tier, the report defines a core set of attributes or distinct organizational capabilities to facilitate the identification and assessment of gaps in coverage in the ICT SCRM community.

CSD also awarded a grant to UMD's Supply Chain Management Center to develop a prototype web portal, which will include:

* Enterprise Risk Assessments: Three-tier risk score-carding system based on the ICT SCRM Community Framework Reference Architecture: Strategic Assessment/ Organizational Readiness, Practices from

NISTIR 7622, and composite network vulnerability map of physical and cyber hubs and nodes, with CVSS ratings;

* Collaboration / Crowdsourcing: User-documented ICT SCRM use/abuse cases and real-time polling about vulnerabilities and responses;

* ICT SCRM Initiatives: A dynamic matrix of current industry and public sector ICT SCRM best practices, standards, and policy reform initiatives which can be updated by appropriate individuals from across industry, academia, and government; and

* ICT SCRM Digital Library: An online repository of policy and academic documents related to ICT SCRM.

In FY2013, CSD will continue to research and develop tools and guidance to help agencies more effectively manage their ICT supply chain risk. In October 2013, CSD intends to publish final NISTIR 7622, *Notional Supply Chain Risk Management Practices for Federal Information Systems*.

CSD also plans to host a workshop in October 2012 to bring together a varied group of stakeholders and thought leaders from industry, academia and government to future NIST/CSD efforts relating to ICT SCRM. Outputs from this workshop will include:

* Fundamental underpinnings of ICT SCRM (terms, definitions, characterizations);

* Current and needed commercially reasonable ICT SCRM-related standards and practices (need, scope, and development approach);

* Current and needed ICT SCRM tools, technology and techniques useful in securing the ICT supply chain; and

* Current and needed research and resources.

CSD plans to use the proceedings of this workshop to transition into producing a draft Special Publication (SP) 800-161 to provide federal departments and agencies guidance on managing the risk to their ICT supply chains. In addition, a number of small forums with academia, industry, and government stakeholders will also be used to inform the development of draft SP 800-161.

Contacts:

Mr. Jon Boyens
ICT SCRM Project Lead
(301) 975-5549
jon.boyens@nist.gov

Ms. Celia Paulsen
(301) 975-5981
celia.paulsen@nist.gov

Nationwide Public Safety Broadband Network (NPSBN) Security

In February 2012, Congress passed the Middle Class Tax Relief and Job Creation Act. One portion of this legislation calls for the establishment of a nationwide, interoperable public safety broadband network based on Long-Term Evolution (LTE) technology to be deployed and operated by the First Responder Network Authority (FirstNet). The planned National Public Safety Broadband Network (NPSBN) is intended to *"create a much-needed nationwide interoperable broadband network that will help police, firefighters, emergency medical service professionals and other public safety officials stay safe and do their jobs."* (http://www.ntia.doc.gov/category/public-safety). In addition, the Director of NIST is called upon to establish a list of certified devices and components to be adhered to when interacting with the nationwide network for the use of public safety officials, vendors, and other interested parties; and to conduct research and development that supports the acceleration and advancement of the nationwide network.

Image Source:
http //www.pscr.gov/index php

In FY2012, CSD advised the National Telecommunications and Information Administration (NTIA) and NIST's Engineering Laboratory on the security of the proposed NPSBN.

NTIA is helping the National Public Safety Telecommunications Council (NPSTC) develop a Statement of Requirements (SoR) to submit to FirstNet that will describe, in increasing levels of detail, the technical requirements of the NPSBN infrastructure, equipment, and communications. This undertaking has required the involvement of representatives of numerous professional communities and a wide variety of disciplines, including federal, state, local, and tribal officials; public safety and other First Responder personnel; mobile carrier operators; and mobile device vendors. CSD commented and contributed text for the security-related aspects of the SoR.

In FY2013, CSD plans to continue supporting the development and refinement of the NPSTC SoR and represent public safety in international standardization efforts, such as the Internet Engineering Task Force (IETF) and the 3rd Generation Partnership Project (3GPP), that impact their communication requirements. In addition, CSD will identify gaps between NIST security guidelines and NPSBN security requirements as well as relevant 3GPP LTE specifications. CSD also plans to research some of the currently unresolved issues and risks related to public safety security.

Contacts:

Ms. Sheila Frankel
(301) 975-3297
sheila.frankel@nist.gov

Dr. Nelson Hastings
(301) 975-5237
nelson.hastings@nist.gov

Smart Grid Cybersecurity

The major elements of the Smart Grid are the information technology, the industrial control systems, and the communications infrastructure used to send command information across the electric grid from generation to distribution systems, and to exchange usage and billing information between utilities and their customers. Key to the successful deployment of the Smart Grid infrastructure is the development of the cybersecurity strategy for the Smart Grid. In fact, cybersecurity needs to be designed into the deploying systems that support Smart Grid, and added into existing systems. The electric grid is critical to the economic and physical well-being of the nation, and emerging cyber threats targeting power systems highlight the need to integrate advanced security to protect critical assets.

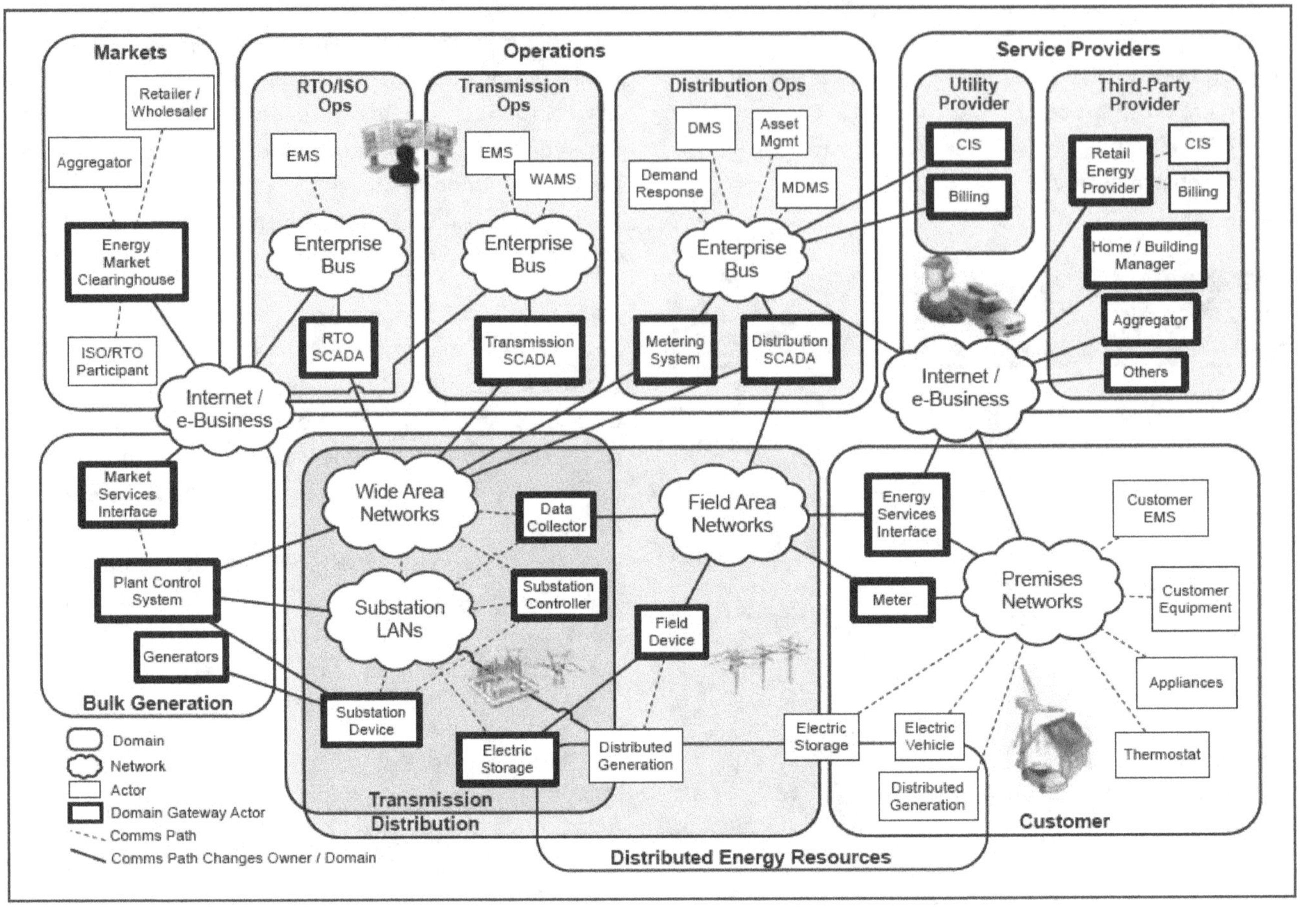

NIST Conceptual Model of the Smart Grid[1]

NIST established the Smart Grid Interoperability Panel (SGIP) Cyber Security Working Group (CSWG) in support of the Energy Independence and Security Act of 2007 to address the cross-cutting issue of cybersecurity. The CSWG has more than 750 participants worldwide from the private sector (including utilities, vendors, and service providers), academia, regulatory organizations, state and local government, and U.S. federal agencies. Membership in the CSWG has been free and open to all since its inception. Many members participate from around the world by monitoring the minutes and email conversations of the subgroups.

The CSWG membership collaborated to deliver the NIST Interagency Report (NISTIR) 7628, *Guidelines for Smart Grid Cyber Security*, in August 2010. Since then the group has focused on specific topics such as risk management processes, key management in the Smart Grid, the Smart Grid security architecture, security testing and certification, Advanced Metering Infrastructure (AMI) security, and privacy in the Smart Grid. In addition, the group is conducting security reviews of many Smart Grid-related standards and beginning to develop a User's Guide for NISTIR 7628.

To complete the work, there are seven subgroups that focus on specific topics. During the development of NISTIR 7628, the subgroups performed detailed technical analyses on an array of security-related topics, and then documented the research, issues, and guidance in specific sections. The approach that has been taken by all subgroups is an open and collaborative process in which any CSWG member is welcome to participate and contribute.

[1]NIST Special Publication 1108R2, NIST Framework and Roadmap for Smart Grid Interoperability Standards, Release 2.0 Available at http://www.nist.gov/smartgrid/upload/NIST_Framework_Release_2-0_corr.pdf.

The CSWG creates and disbands subgroups in order to meet present needs. Since the NISTIR 7628 publication, some of the CSWG subgroups have merged, while others are regrouping as they determine their next set of tasks. The CSWG currently consists of the following subgroups:

* The **Architecture subgroup** focuses on the enhancement of the logical security architecture for the Smart Grid. This group's work is used as input to the SGIP Architecture Committee;

* The **High-Level Requirements subgroup** addresses the procedural and technical security requirements of the Smart Grid to be addressed by stakeholders in Smart Grid security. To create the initial set of security requirements in NISTIR 7628, this subgroup adapted industry-accepted security source documents for the Smart Grid;

* The **NISTIR 7628 User's Guide subgroup** will provide an easy-to-understand tool that utilities and other entities involved in implementing Smart Grid-based systems can use to navigate NISTIR 7628 to identify and select the security requirements needed to help protect those systems;

* The **Privacy subgroup** continues to investigate privacy concerns between utilities, consumers, and nonutility third parties;

* The **Standards subgroup** assesses standards and other documents with respect to the cybersecurity and privacy requirements from NISTIR 7628. These assessments are performed on the standards contained in NIST Special Publication (SP) 1108, *Framework and Roadmap for Smart Grid Interoperability Standards*, or in support of the SGIP's Priority Action Plans (PAPs). The group has reviewed over 75 documents to date; and

* The **Testing and Certification subgroup** establishes guidance and methodologies for cybersecurity testing of Smart Grid systems, subsystems, and components.

During the past year, members of the CSWG collaborated with the Department of Energy (DOE) and the North American Electric Reliability Corporation (NERC) to develop a harmonized electricity sector enterprise-wide risk management process. This was published by DOE as "Electricity Subsector Cybersecurity Risk Management Process Guideline" in May 2012. The CSWG also collaborated with the National Electric Sector Cybersecurity Organization Resource (NESCOR) to develop a technical white paper on Smart Energy Profile (SEP) 1.0 and 1.1. The CSWG Lead, Marianne Swanson, also served as a subject-matter expert on the DOE Electricity Subsector Cybersecurity Capability Maturity Model (ES-C2M2) project.

Members of the CSWG produced a white paper on automating Smart Grid security in December 2011. The group produced a mapping between the NERC Critical Infrastructure Protection standards, version 5, and the high-level security requirements found in NISTIR 7628. For the SGIP's second version of the *Interoperability Process Reference Manual*, the group produced a chapter on cybersecurity testing. The group also developed an SGIP document, *Guide for Testing the NISTIR 7628 High-level Security Requirements*, in order to help utilities and other Smart Grid organizations assess how well they are meeting those high-level security requirements.

The CSWG's Privacy subgroup developed customizable train-the-trainer privacy briefings for utilities, public utility commissions, and those dealing with consumers. This subgroup also developed a set of recommended practices for how third parties should protect consumers' privacy when handling customer energy usage data received from a source other than a utility (e.g., an in-home device).

CSD also supports CSWG by assessing the security of cryptographic methods used in security protocols for utilities' communication and management networks and by developing cryptographic security guidelines for the Smart Grid. During the past year, CSD reviewed and analyzed cryptographic techniques in the Smart Energy Profile (SEP) standards, and provided guidance to the CSWG. The EAX' block cipher mode of operation, specified in the ANSI C12.22-2008 standard, was intended by industry to serve as a solution for cybersecurity needs in smart meters. CSD reviewed the EAX' scheme, submitted by members of the ANSI C11.22 SC 17 Committee, and decided not to include it in the toolkit of NIST-approved algorithms. CSD also reviewed the IEC 62541-6 standard (*OPC unified architecture - Part 6: Mappings*)[2] and identified issues with the cryptographic methods used in the standard that could potentially create a substantial security risk for communication networks running the standard protocol. CSD provided recommendations to improve the security and performance of this standard.

CSD also represented the CSWG, in collaboration with the Electric Power Research Institute (EPRI) and Department of Energy/NESCOR, in identifying requirements, features, and capabilities for a cryptographic key-management system for an Advanced Metering Infrastructure (AMI). This is an ongoing work project.

Future work includes working with the SGIP — the Committees, the Domain Expert Working Groups, and the Priority Action Plans — to integrate cybersecurity into their work efforts. Collaboration will continue with DOE and NERC to produce a cybersecurity risk management process use case for the electricity sector to accompany the Risk Management Process document. Reviewing and updating NISTIR 7628 will occur in the next year. This will result in the issuance of NISTIR 7628 Revision 1. Afterwards, the group will update the *Guide for Testing the NISTIR 7628 High-level Security Requirements* to reflect any changes in the high-level security requirements from NISTIR 7628 Rev. 1. Members of the group will produce a white paper on cloud computing and the Smart Grid, as well as the NISTIR 7628 User's Guide. The Standards subgroup will continue to review documents for the SGIP. CSD will continue to support the CSWG in the evaluation of the cryptographic methods used in security protocols for the communication and management networks used by utility companies.

http://collaborate.nist.gov/twiki-sggrid/bin/view/SmartGrid/WebHome

Contacts:

Ms. Marianne Swanson
(301) 975-3293
marianne.swanson@nist.gov

Ms. Tanya Brewer
(301) 975-4534
tbrewer@nist.gov

Mr. Quynh Dang
(301) 975-3610
qdang@nist.gov

Cybersecurity Awareness, Training, Education, and Outreach

→ National Initiative for Cybersecurity Education (NICE)

NIST became the lead for the National Initiative for Cybersecurity Education (NICE) in 2010 based on a recommendation of the Information and Communications Infrastructure - Interagency Policy Committee (ICI-IPC). This recommendation was built out of chapter two of the May 2009 Cyberspace Policy Review titled "Building Capacity for a Digital Nation" and is responsive to President Obama's declaration that the "cyber threat is one of the most serious economic and national security challenges we face as a nation" and that "America's economic prosperity in the 21st century will depend on cybersecurity."

The goal of NICE is to enhance the overall cybersecurity posture of the United States by accelerating the availability of educational and training resources designed to improve the cyber behavior, skills, and knowledge of every segment of the population, enabling a safer cyberspace for all. NICE addresses this challenging goal by:

* Raising national awareness about risks in cyberspace;
* Broadening the pool of individuals prepared to enter the cybersecurity workforce; and
* Cultivating a globally competitive cybersecurity workforce.

---

[2]Object Linking and Embedding (OLE) for Process Control

This initiative comprises four component areas: 1) National Cybersecurity Awareness; 2) Formal Cybersecurity Education; 3) Cybersecurity Workforce Structure; and 4) Cybersecurity Workforce Training and Professional Development.

The Computer Security Division (CSD) is home to the NIST NICE Leadership Team (NNLT), and they focus on the following activities for NICE:

* Developing planning documents, and building consensus on the strategy and implementation activities of NICE;
* Facilitating cross-functional cooperation among NICE component lead agencies;
* Fostering communication between the component lead agencies by coordinating meetings, facilitating discussions, and disseminating information;
* Promoting the initiative and its efforts by representing NICE and speaking at cybersecurity events nationwide;
* Planning and hosting an annual workshop to promote and support the evolving issues in cybersecurity education;
* Coordinating with other federal initiatives and efforts related to NICE; and
* Maintaining and updating the NICE website.

In FY2012, NIST updated the draft NICE Strategic Plan and stewarded the National Cybersecurity Workforce Framework, developed within NICE's Cybersecurity Workforce Training and Professional Development Component, through government-wide review. Both documents can be found on the NICE website, http://csrc.nist.gov/nice/.

NIST organized and planned to host the third annual NICE Workshop, *"Shaping the Future of Cybersecurity Education, Connecting the Dots in Cyberspace,"* from October 30 - November 1, 2012. However, the workshop was cancelled due to Hurricane Sandy. The next NICE workshop will be held in the fall of 2013 and will serve as a forum to develop the connections within the cybersecurity and education communities to make progress on the strategic goals and objectives of NICE.

The NNLT attended more than 100 events, symposia, forums, competitions, educational outreach meetings, and workshops to promote the activities within NICE. In FY2012, the National Cybersecurity Education Council was established with the signing of a Memorandum of Understanding between the Department of Education, NIST, and over 50 private sector organizations convened by the National Cyber Security Alliance.

In FY2013, the NNLT plans to develop innovative strategies to support the Office of Personnel Management (OPM) issuance and the governmentwide adoption of cybersecurity function codes that will, for the first time, allow agencies to identify their cybersecurity workforce. The NNLT will also demonstrate the value of framework adoption to government agencies and private sector organizations who are formalizing their cybersecurity training needs. The NNLT will be identifying opportunities to tie the framework to existing and new training initiatives.

The NNLT has cultivated the trust of the Component lead agencies. In FY2013, NNLT will use that trust to best support the Department of Homeland Security (DHS) development of the National Initiative for Cybersecurity Careers and Studies (NICCS) web portal. All of the Components have identified deliverables and milestones, and the NNLT will work to publicize the accomplishments of the Components.

http://www.nist.gov/nice/

Contacts:

Dr. Ernest McDuffie
NICE Project Lead
(301) 975-8897
ernest.mcduffie@nist.gov

Mr. Bill Newhouse
NICE Program Lead
(301) 975-2869
william.newhouse@nist.gov

Ms. Magdalena Benitez
(301) 975-6182
mbenitez@nist.gov

Ms. Pat Toth
(301) 975-5140
ptoth@nist.gov

→ Computer Security Resource Center (CSRC)

The Computer Security Resource Center (CSRC), CSD's website, is one of the most visited websites at NIST. CSRC encourages broad sharing of information security tools and practices, provides a resource for information security standards and guidelines, and identifies and links key security web resources to support industry and government users. CSRC is an integral component of all of the work that CSD conducts and produces. It is CSD's repository for anyone wanting to access these documents and other valuable security-related information. During FY2012, CSRC had more than 51 million requests.[1]

TOTAL NUMBER OF WEBSITE REQUESTS: CSRC

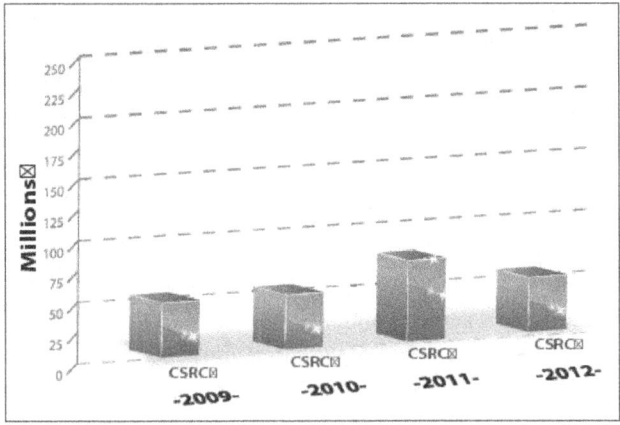

CSRC is the primary gateway for gaining access to NIST computer security publications, standards, and guidelines, and serves as a vital link to CSD's customers.

The URL for the Publications homepage is: http://csrc.nist.gov/publications. Publications are organized by Topic, Family categories, and Legal Requirements to help users locate relevant information quickly.

During FY2012, the top ten most popular publications were:

(1) SP 800-53 Revision 3, *Recommended Security Controls for Federal Information Systems and Organizations*;

(2) SP 800-12, *An Introduction to Computer Security: The NIST Handbook*;

(3) Draft SP 800-53 Revision 4, *Security and Privacy Controls for Federal Information Systems and Organizations*;

(4) FIPS 140-2, *Security Requirements for Cryptographic Modules*;

(5) SP 800-94, *Guide to Intrusion Detection and Prevention Systems (IDPS)*;

(6) SP 800-77, *Guide to IPsec VPNs*;

(7) SP 800-61 Revision 2, *Computer Security Incident Handling Guide*;

(8) SP 800-100, *Information Security Handbook: A Guide for Managers*;

(9) SP 800-30, *Guide for Conducting Risk Assessments*; and

(10) SP 800-64 Revision 2, *Security Considerations in the System Development Life Cycle*.

In addition to CSRC, CSD maintains a publication announcement mailing list. This free email list notifies subscribers about publications that have been posted to the CSRC website. The email list is a valuable tool for more than 20,000 subscribers from the federal government, industry, and academia, and individuals with a personal interest in IT security. Individuals who are interested in subscribing to this list should visit http://csrc.nist.gov/publications/subscribe.html for more information.

Questions on the website should be sent to the CSRC Webmaster at: webmaster-csrc@nist.gov.

Contacts:

Mr. Patrick O'Reilly
(301) 975-4751
patrick.oreilly@nist.gov

Ms. Judy Barnard
(301) 975-5502
jbarnard@nist.gov

---

[1] These statistics are based from October 1, 2011 to September 30, 2012 time frame. The total requests consist of web pages and file downloads.

→ Federal Computer Security Program Managers' Forum

The Federal Computer Security Program Managers' Forum is a group that is sponsored by NIST to promote the sharing of security-related information among federal agencies. The Forum, which serves more than 1,100 members, strives to provide an ongoing opportunity for managers of federal information security programs to exchange information security materials in a timely manner, build upon the experiences of other programs, and reduce possible duplication of effort. It provides a mechanism for NIST to share information directly with federal agency information security program managers in fulfillment of NIST's leadership mandate under FISMA. It also assists NIST in establishing and maintaining relationships with other individuals or organizations that are actively addressing information security issues within the federal government. NIST serves as the Secretariat of the Forum, providing necessary administrative and logistical support. Kevin Stine serves as the Chairperson for the Forum. Participation in Forum meetings is open to federal government employees who participate in the management of their organization's information security program as well as their designated support contractors. There are no membership dues.

The Forum maintains an extensive email list, holds bimonthly meetings and an annual two-day conference to discuss current issues and developments of interest to those responsible for protecting sensitive (unclassified) federal systems. The Forum plays a valuable role in helping NIST and other federal agencies to develop and maintain a strong, proactive stance in the identification and resolution of new strategic and tactical IT security issues as they emerge.

Topics of discussion at Forum meetings in FY2012 included briefings from various federal agencies on a telework reference architecture; National Archives and Records Administration's (NARA) Controlled Unclassified Information (CUI) Implementation Guidance for Executive Order 13556; Federal Public Key Infrastructure (PKI) Security Profile; Combinatorial Methods in Software Testing; NIST SP 800-63-1, *Electronic Authentication Guide*; Federal Risk and Authorization Management Program (FedRAMP); Treasury's PIV Program Implementation; and Electricity Subsector Cybersecurity Risk Management.

This year's annual two-day offsite meeting featured updates on the computer security activities of the Government Accountability Office (GAO), United States Computer Emergency Readiness Team (US-CERT), the National Security Staff, and NIST. Technical sessions included briefings on evolving cybersecurity strategies, Basic Input/Output System (BIOS) security mechanisms, the Cryptographic Module Validation Program (CMVP), Supply Chain Risk Management, Big Data, the National Initiative for Cybersecurity Education (NICE), Transportation's privacy program implementation, and a General Services Administration (GSA) Cloud implementation case study.

The number of members on the email list has grown steadily and provides a valuable resource for federal security program managers. To join, email your name, affiliation, address, phone number, title, and confirmation that you are a federal employee to sec-forum@nist.gov.

http://csrc.nist.gov/groups/SMA/forum/

Contacts:

Mr. Kevin Stine	Ms. Peggy Himes
Chair	Administration
(301) 975-4483	(301) 975-2489
kevin.stine@nist.gov	peggy.himes@nist.gov

→ Federal Information Systems Security Educators' Association (FISSEA)

The Federal Information Systems Security Educators' Association (FISSEA), founded in 1987, is an organization run by and for information systems security professionals to assist federal agencies in meeting their information systems security awareness, training, and education responsibilities. FISSEA strives to elevate the general level of information systems security knowledge for the federal government and the federal workforce. FISSEA serves as a professional forum for the exchange of information and improvement of information systems security awareness, training, and education programs. It also seeks to assist the professional development of its members.

During the 2011 conference business meeting, it was announced that NIST would make a deeper commitment to FISSEA. NIST's plan includes a graceful transition to a NIST program supported by the current Executive Board, and include direct and formal connections with the National Initiative for Cybersecurity Education (NICE).

FISSEA membership is open to information systems security professionals, professional trainers and educators, and managers responsible for information systems security training programs in federal agencies, as well as contractors of these agencies and faculty members of accredited educational institutions who are involved in information security training and education. There are no membership fees to join FISSEA; all that is required is a willingness to share products, information, and experiences. Business is administered by a working group that meets monthly.

FISSEA maintains a website, a mailing list, and participates in a social networking site as a means of improving communication for its members. NIST assists FISSEA with its operations by providing staff support for several of its activities and by being FISSEA's host agency.

FISSEA membership in 2012 spanned federal agencies, industry, military, contractors, state governments, academia, the press, and foreign organizations to reach over 1,295 members in a total of ten countries. The 700 federal agency members represent 89 agencies from the executive and legislative branches of government.

The 2012 FISSEA conference returned to NIST on March 27-29, 2012, and the theme was "A New Era in Cybersecurity Awareness, Training, and Education." The theme was chosen to reflect current projects, trends, and initiatives that provide pathways to future solutions. Approximately 184 information systems security professionals and trainers from federal agencies, academia, and industry attended. Attendees received an update on NICE activities, gained new techniques for developing and conducting training, as well as awareness and training ideas, resources, and contacts. Presenters represented NIST, the Department of Homeland Security (DHS), the Defense Intelligence

Agency (DIA), the U.S. Department of State (DOS), the Department of Energy (DOE), the Department of Defense (DoD), the Department of Veterans Affairs (VA), the National Aeronautics and Space Administration (NASA), the National Security Agency (NSA), the Department of Interior, the Bureau of the Public Debt (BPD), and the Library of Congress. Presenters also represented private industry and academia. Attendees had an opportunity to visit 22 vendors on the second day. Another bonus of attending the FISSEA conference is networking.

The conference continues to be a valuable forum in which individuals from government, industry, and academia involved with information systems/cybersecurity workforce development - awareness, training, education, certification, and professionalization - may learn of ongoing and planned training and education programs and initiatives.

At each annual conference, an award is presented to a candidate selected as FISSEA Educator of the Year; this award honors distinguished accomplishments in information systems security training programs. Susan Hansche, Avaya Government Solutions/U.S. Department of State was awarded the Educator of the Year for 2011 at the 2012 FISSEA Conference. The annual FISSEA Security Awareness, Training and Education Contest consists of five categories from one of FISSEA's three key areas of Awareness, Training, and Education. The categories are: (1) awareness poster, (2) motivational item (aka: trinkets - pens, stress relief items, t-shirts, etc.), (3) awareness website, (4) awareness newsletter, and (5) role-based training and education. Winning entries for the security awareness contest are posted to the FISSEA website. The winners for the FY2012 contest were:

* David Kurtz and Bruce Sharp, Bureau of the Public Debt, U.S. Treasury Department won the Poster Contest;
* Maureen Moore, Sara Fitzgerald, Kimberly Conway, and Mechelle Munn, Food and Drug Administration, was selected as winners for their Security Motivation Item as well as for their Security Newsletter; and

* Shelly Tzoumas, U.S. House of Representatives, had the winning security website and was selected as the Role-Based Training Contest winner.

New in FY 2012 was a poster session which provided an opportunity to share and tell about their specific awareness and training programs. During the poster session, conference attendees voted for their favorite selection in each category to select the "Peer's Choice" award. Attendees selected the same newsletter, motivational item, and security training winners as the contest judging committee. However, peers selected the poster submitted by Terri Cinnamon, Department of Veterans Affairs, and Alexis Benjamin, Department of State, as the peer's choice for website.

The 2013 FISSEA conference is being planned for March 19-21, 2013, at NIST.

http://csrc.nist.gov/fissea
fisseamembership@nist.gov

Contacts:

Ms. Patricia Toth
(301) 975-5140
patricia.toth@nist.gov

Ms. Peggy Himes
(301) 975-2489
peggy.himes@nist.gov

→ Information Security and Privacy Advisory Board (ISPAB)

The Information Security and Privacy Advisory Board (ISPAB) is a federal advisory committee. It brings together senior professionals from industry, government, and academia to advise NIST, the Office of Management and Budget (OMB), the Secretary of Commerce, and appropriate committees of the U.S. Congress about information security and privacy issues pertaining to unclassified federal government information systems.

In accordance with 15 U.S.C. 278g-4, the ISPAB is rechartered by the Secretary of Commerce for 2012-2013 in pursuant to the Federal Advisory Committee Act, 5 U.S.C. App. The scope and objectives of the Board are to—

* Identify emerging managerial, technical, administrative, and physical safeguard issues relative to information security and privacy;

* Advise NIST, the Secretary of Commerce, and the Director of OMB on information security and privacy issues pertaining to federal government information systems, including thorough review of proposed standards and guidelines developed by NIST; and

* Annually submit a report to the Secretary of Commerce, the Director of OMB, the Director of the National Security Agency, and the appropriate committees of the Congress.

The charter (http://csrc.nist.gov/groups/SMA/ispab/documents/ispab_charter-2012-2014.pdf) defines that the Board's membership consists of twelve members and a Chairperson. The Chairperson is appointed by the Secretary of Commerce, and the Board members are selected for their preeminence in the information technology industry or related disciplines. The term of office for each board member is four years. The ISPAB Board members are:

* Daniel Chenok (Chair), IBM Center for The Business of Government;

* Julie Boughn, Center for Medicare and Medicaid Innovation, Department of Human Health and Services, Centers for Medicare & Medicaid Services (DHHS/CMS);

* Christopher Boyer, AT&T;

* Kevin Fu, University of Massachusetts Amherst;

* Gregory Garcia, Garcia Cyber Partners;

* Brian Gouker, National Security Agency (NSA) - U.S. Army War College;

* Toby Levin, (retired);

* Edward Roback, U.S. Department of the Treasury;

* Phyllis Schneck, McAfee, Inc.;

* Gale Stone, Social Security Administration;

* Matthew Thomlinson, Microsoft; and

* Peter Weinberger, Google, Inc.

This advisory board of experienced, dynamic, and knowledgeable professionals provides NIST and the federal government with a rich, varied pool of people conversant with an extraordinary range of topics.

Left to Right: Annie Sokol, Chris Boyer, Kevin Fu, Toby Levin, Ed Roback, Greg Garcia, Phyllis Schneck, Dr. Pat Gallagher (NIST Director), Peter Weinberger, Charles Romine (ITL Director), and Dan Chenok (Chair, ISPAB)

The Board's membership draws from experience at all levels of information security and privacy work. The members' careers cover government, industry, and academia. Members have worked in the executive and legislative branches of the federal government, civil service, senior executive service, the military, some of the largest corporations worldwide, small and medium-size businesses, and some of the top universities in the nation. The members' experience, likewise, covers a broad spectrum of activities including many different engineering disciplines, computer programming, systems analysis, mathematics, management, information technology auditing, privacy, and law. Members also have an extensive history of professional publications and professional journalism. Members have worked (and in many cases, continue to work) on the development and evolution of some of the most important pieces of information security and privacy legislation in the federal government, including the Privacy Act of 1974, the Computer Security Act of 1987, the E-Government Act (including FISMA), and other e-government services and initiatives.

In FY2012, the board lost a valuable member, Joseph Guirreri, who had diligently served the Board for the past six years. In the same year, F. Lynn McNulty, a board member (2005-2011) and a significant contributor to information security in the government, passed away on June 4th.

The Board usually meets three times per year and meetings are open to the public. NIST provides the Board with its Secretariat and meetings are usually located in Washington, D.C. In June 2012, NIST requested to host ISPAB at NIST and showcased a range of research for the Board's discussion. The Board has received numerous briefings from federal and private sector representatives on a wide range of privacy and security topics in the past year. Areas of interest that the Board followed in FY2011-FY2012 were:

* Special Publication (SP) 800-53 Revision 4, *Security and Privacy Controls for Federal Information Systems and Organizations*, and its appendices;
* Federal Communications Commission (FCC) and technology;

* Cyber defense;
* Prospects for and content of cyber legislature;
* Consumer privacy;
* Security in the next generation mobility;
* Cloud Computing - data location, data storage, and data sovereignty;
* Legislature and security;
* Derived credentials;
* Economic incentives for medical device cybersecurity;
* Cyber ecosystem and automated cyber indicator sharing;
* Key Management;
* Defense Industrial Base (DIB) pilot and its potential application to other private sectors;
* Progress for modernizing federal desktop platforms;
* Trustworthy cyberspace: Strategic Plan for the Federal Cybersecurity Research and Development Program; and
* Federal initiatives such as:
    ◦ National Initiative for Cybersecurity Education (NICE);
    ◦ National Strategy for Trusted Identities in Cyberspace (NSTIC);
    ◦ Federal Risk and Authorization Management Pilot program (FedRAMP);
    ◦ United States Computer Emergency Readiness Team (US-CERT);
    ◦ Homeland Security Presidential Directive (HSPD) 12;
    ◦ Draft FIPS 201-2, *Personal Identity Verification (PIV) of Federal Employees and Contractors*;
    ◦ National Cybersecurity and Communications Integration Center (NCCIC) and Cyber Storm - Automated Indicator Sharing;
    ◦ Continuous Monitoring;
    ◦ IT System Performance and Conformity;
    ◦ Federal Guide to Privacy and Security of Health;

◦ Securities and Exchange Commission (SEC) Security Breach Notification;
◦ Federal Information Security Management Act (FISMA); and
◦ NIST's outreach, research, and strategies.

http://csrc.nist.gov/groups/SMA/ispab/index.html

Contact:
Ms. Annie Sokol
(301) 975-2006
annie.sokol@nist.gov

→ Small and Medium-Size Business (SMB) Outreach

What do business invoices have in common with email? If both are done on the same computer, the business owner may want to think more about computer security information - payroll records, proprietary information, client or employee data - as essential to a business's success. A computer failure or system breach could cost a business anything from its reputation to damages and recovery costs. The small business owner who recognizes the threat of computer crime and takes steps to deter inappropriate activities is less likely to become a victim.

The vulnerability of any one small business may not seem significant to many people, other than the owner and employees of that business. However, over 25 million U.S. businesses, comprising more than 99 percent of all U.S. businesses, are small and medium-size businesses (SMBs) of 500 employees or less (http://www.sba.gov/sites/default/files/us11_0.pdf). Therefore, a vulnerability common to a large percentage of SMBs could pose a threat to the nation's information infrastructure and economic base. SMBs frequently cannot justify the employment of an extensive security program or a full-time expert. Nonetheless, they confront serious security challenges.

The difficulty for these businesses is to identify security mechanisms and training that are practical and cost-effective. Such businesses also need to become more educated in terms of security so that limited resources are well applied to meet the most relevant and serious threats. To address this need, NIST, the Small Business Administration (SBA), and the Federal Bureau of Investigation (FBI) are cosponsoring a series of training workshops on computer security for small businesses. The purpose of the workshops is to provide an overview of information security threats, vulnerabilities, and corresponding protective tools and techniques, with a special emphasis on providing useful information that small business personnel can apply directly.

In FY2012, twenty-five SMB outreach workshops were provided in twenty-four cities: Tulsa, Oklahoma; Oklahoma City, Oklahoma; Lake Charles, Louisiana; Lafayette, Louisiana; Baton Rouge, Louisiana; Slidell, Louisiana; New Orleans, Louisiana; Denver, Colorado; Nashua, New Hampshire; New Haven, Connecticut; Rochester, Minnesota; St Paul, Minnesota; St Cloud, Minnesota; Austin, Texas; San Antonio, Texas; San Diego, California; Los Angeles, California; Albuquerque, New Mexico; Indianapolis, Indiana; Sacramento, California; Cincinnati, Ohio; Dayton, Ohio; Columbus, Ohio; and Chillicothe, Ohio.

In collaboration with the SBA and the FBI, planning is under way to identify locations for small business information security workshops in FY2013.

http://sbc.nist.gov

Contact:
Mr. Richard Kissel
(301) 975-5017
richard.kissel@nist.gov

→ **Crypto Standards Program**

Hash Algorithms and the Secure Hash Algorithm (SHA-3) Competition

The Cryptographic Technology Group (CTG) is responsible for the maintenance and development of the *Secure Hash Standard (SHS)*, FIPS 180. A hash algorithm processes a message, which can be very large, and produces a condensed representation of the message, called a message digest. A cryptographic hash algorithm is a fundamental component of many cryptographic functions, such as digital signature algorithms, key-derivation functions, keyed-hash message authentication codes (HMAC), or random number generators. Cryptographic hash algorithms are frequently used in Internet protocols or in other security applications.

In 2005, researchers developed an attack method that threatened the security of the Secure Hash Algorithm-1 (SHA-1), a NIST-approved hash algorithm. Researchers at NIST and elsewhere also discovered several generic limitations in the basic Merkle-Damgard construct that is used in SHA-1 and most other existing hash algorithms. To address these vulnerabilities, NIST opened a public competition in November 2007 to develop a new cryptographic hash algorithm, which would be called "SHA-3," and would augment the Secure Hash Standard by adding the new hash algorithm.

NIST received sixty-four entries from cryptographers around the world by October 31, 2008, and selected fifty-one first-round candidates in December 2008; fourteen second-round candidates in July 2009; and five third-round candidates - BLAKE, Grøstl, JH, Keccak and Skein, on December 9, 2010, to enter the third and final round of the competition. Status reports for the first and second rounds were published as NISTIR 7620, *Status Report on the First Round of the SHA-3 Cryptographic Hash Algorithm Competition*, and NISTIR 7764, *Status Report on the Second Round of the SHA-3 Cryptographic Hash Algorithm Competition*, respectively.

Submitters of the SHA-3 finalists were allowed to make minor adjustments to their algorithms by January 16, 2011, and the third round of the competition began on January 31, 2011, when the final submissions were posted on NIST's hash website. A one-year public review period was provided before NIST hosted the (last) Third SHA-3 Candidate Conference (URL: http://csrc.nist.gov/groups/ST/hash/sha-3/Round3/March2012/) in Washington, D.C. on March 22-23, 2012, to receive public feedback on the finalists.

The cryptographic community provided an enormous amount of expert feedback throughout the competition. Most of the comments were sent to NIST and a public hash forum; in addition, many of the cryptanalysis and performance studies were published as papers in major cryptographic conferences or leading cryptographic journals. Based on the public comments and internal review of the candidates, NIST announced KECCAK as the SHA-3 winner on October 2, 2012, thus ending the five-year-long competition.

KECCAK was selected because of its large security margin, good general performance, excellent efficiency in hardware implementations, flexible design, and because its design and implementation properties complement the existing SHA-2 family of hash algorithms well. The evaluation of the finalists and the selection process was summarized in a third-round report, which was published as NISTIR 7896 in early FY2013.

NIST plans to augment the current hash standard, FIPS 180-4, to include the new SHA-3 algorithm, and publish a draft FIPS 180-5 for public review. After the close of the public comment period, NIST will revise the draft standard, as appropriate, in response to the public comments that NIST receives. A final review, approval, and promulgation process will then follow.

http://www.nist.gov/hash-competition

Contact:
Ms. Shu-jen Chang
(301) 975-2940
shu-jen.chang@nist.gov

Hash Algorithm Standards and Security Guidelines

FIPS 180 is the SHA standard. This standard has had several revisions. FIPS 180-3 was approved in October 2008 and contained five hash algorithms: SHA-1, SHA-224, SHA-256, SHA-384 and SHA-512.

In March 2012, another revision of FIPS 180 was approved: FIPS 180-4. This revision provides a general procedure for creating an initialization hash value, adds two additional secure hash algorithms (SHA-512/224 and SHA-512/256) to the SHA standard, and removes a restriction that padding must be done before hash computation begins, which was required in FIPS 180-3. SHA-512/224 and SHA-512/256 are more efficient alternatives to SHA-224 and SHA-256 on platforms that are optimized for 64-bit operations. Removing the restriction on the padding operation in the secure hash algorithms potentially allowed more flexibility and efficiency in implementing the secure hash algorithms in many computer network applications.

General guidelines for using hash functions are provided in NIST Special Publication (SP) 800-107, Revision 1, *Recommendation for Applications Using Approved Hash Algorithms*. This document provides security guidelines for achieving the desired security strengths for cryptographic applications that employ the approved cryptographic hash functions specified in FIPS 180. SP 800-107 has been revised to address the security properties of SHA-512/224 and SHA-512/256, the new hash algorithms approved in FIPS 180-4. Additional security information about Hash Message Authentication Code (HMAC) has been provided, and the hash-based key-derivation function section has been revised to provide updated information about approved hash-based key-derivation functions that are specified in many other NIST Special Publications.

A draft of the SP 800-107 revision was issued for public comment in September 2011 and was extensively revised to address the received comments and to include discussions of the security of HMAC and randomized hashing for digital signatures. The discussions on hash-based key-derivation functions were revised to incorporate the "extraction-then-expansion" key-derivation procedure specified in SP 800-56C, *Recommendation for Key Derivation through Extraction-then-Expansion*, and to discuss different approved hash-based key derivation functions. The revision was completed and published in August 2012.

In 2013, work will begin on a new revision of the standard (FIPS 180-5), which will contain the new hash algorithm resulting from the SHA-3 competition discussed in the *Hash Algorithms and the Secure Hash Algorithm (SHA)-3 Competition* section of this report.

Contacts:

Mr. Quynh Dang	Ms. Elaine Barker
(301)-975-3610	(301)-975-2911
quynh.dang@nist.gov	ebarker@nist.gov

Random Number Generation (RNG)

Random numbers are needed to provide the required security for most cryptographic algorithms. For example, random numbers are used to generate the keys needed for encryption and digital signature applications.

In the late 1990s, a project to develop more rigorous requirements and specifications for random number generation (RNG) was initiated in coordination with the American National Standards Institute's (ANSI) Accredited Standards Committee (ASC) X9. The resulting standard (X9.82) is being developed in four parts: Part 1 provides general information; Part 2 will provide requirements for entropy sources; Part 3 provides specifications for deterministic random bit generator (DRBG) mechanisms; and Part 4 provides guidance on constructing random bit generators (RBGs) from entropy sources and DRBG mechanisms. Parts 1, 3 and 4 have been completed; Part 2 is nearing completion.

In March 2007, NIST published SP 800-90, *Recommendation for Random Number Generation Using Deterministic Random Bit Generators*, which contained the DRBG mechanisms in Part 3 of ANS X9.82, plus an additional DRBG mechanism. This Recommendation was revised as SP 800-90A, *Recommendation for Random Number Generation*

Using Deterministic Random Bit Generators, in January 2012 to include additional capabilities identified during the development of Part 4 of ANS X9.82. The document number for SP 800-90 was modified so that two additional documents (i.e., SP 800-90B, *Recommendation for the Entropy Sources Used for Random Bit Generation* and SP 800-90C, *Recommendation for Random Bit Generator (RBG) Constructions*) could be included in a series on random number generation. SP 800-90A is available at http://csrc.nist.gov/publications/PubsSPs.html.

SP 800-90B will address the development and testing of entropy sources, including descriptions of the validation tests that will be used by NIST's Cryptographic Algorithm Validation Program to validate candidate entropy sources. SP 800-90C will provide basic guidance on the construction of RBGs from entropy sources and DRBG mechanisms, pointing to Part 4 of ANS X9.82 for additional constructions and examples. Both documents have been provided for public comment at http://csrc.nist.gov/publications/PubsDrafts.html.

NIST's standards activities in FY2013 will include a workshop to discuss the drafts of SP 800-90B and C, and adjudication of the comments received during the public-comment period and the workshop.

Contacts:

Ms. Elaine Barker	Dr. John Kelsey
(301) 975-2911	(301) 975-5101
ebarker@nist.gov	john.kelsey@nist.gov

Key Management

NIST continues to address cryptographic key management for the federal government, and to coordinate this guidance with other national and international organizations, industry, and academia. This guidance has been published as NIST Special Publications (SPs), which are available at http://csrc.nist.gov/publications/PubsSPs.html.

SP 800-56A, *Recommendation for Pair-Wise Key Establishment Schemes Using Discrete Logarithm Cryptography*, specifies approved methods for key establishment using Diffie-Hellman and Menezes-Qu-

Vanstone (MQV) schemes. This document, which was first published in 2006, is being revised to provide further clarification and additional methods for key derivation; this revision has been made available for public comment at http://csrc.nist.gov/publications/PubsDrafts.html and will be completed in FY2013.

A newly approved method for key derivation in SP 800-56A is specified in SP 800-56C, *Recommendation for Key Derivation through Extraction-then-Expansion*, which was completed in November 2011. SP 800-56C specifies a two-step key-derivation procedure that extracts randomness from a shared secret produced during a key-agreement computation and expands the result into the required keying material. The procedure in SP 800-56C is one of the new key-derivation methods referenced in the revision of SP 800-56A.

Another related publication, SP 800-135, *Recommendation for Existing Application-Specific Key Derivation Functions*, was completed in December 2010 and revised in December 2011; this document approves existing application-specific key-derivation functions used in commonly deployed protocols. These key-derivation functions are among the new key-derivation methods referenced in the revision of SP 800-56A.

Part 1 of SP 800-57, *Recommendation for Key Management: Part 1: General*, provides general key-management guidance. This document was first published in 2005, and later revised in 2007. This document has been updated to include information on and references to recent work performed by the CTG; the revision was completed in July 2012.

Part 3 of SP 800-57, *Recommendation for Key Management, Part 3 Application-Specific Key Management Guidance*, was first published in 2009; this document provides application-specific key management guidance and is being revised to reflect recent work on the applications and protocols discussed in the document. The revision will also include an additional section on the Secure Shell (SSH) protocol. In FY2013, the revision of this document will be provided for public comment.

SP 800-130, *A Framework for Designing Cryptographic Key Management Systems*, is being developed to provide guidance on the framework of a Cryptographic Key Management System (CKMS). The first draft of this document was provided for public comment in 2010 and was discussed in a subsequent workshop at NIST in late FY2010. A revised draft of this document that addressed the comments received during the public comment period and during the workshop was provided for a second public-comment period in April 2012 and discussed at a public key management workshop in September 2012. This document will be completed by mid FY2013.

SP 800-152, *A Profile for U.S. Federal Cryptographic Key Management Systems (CKMS)*, is under development. This document is intended to provide refinements of the framework requirements in SP 800-130 that are appropriate for use in a CKMS used by the federal government, plus guidance on implementing, procuring, installing, configuring, and operating a Federal CKMS. A table of proposed requirements to be included in SP 800-152 was provided for public comment in August 2012 and discussed at a key management workshop in September 2012. SP 800-152 will continue to be developed during FY2013, including refining the requirements provided for public comment.

SP 800-133, *Recommendation for Cryptographic Key Generation*, which discusses the generation of the keys to be managed and used by NIST's approved cryptographic algorithms, was issued for public comment in August 2011. It addresses the generation of a key using the output of a random bit generator, the derivation of a key from another key, the derivation of a key from a password, and keys generated during the use of a key-agreement scheme. This document was completed in November 2012.

A new publication is under development that is intended to provide guidance on the security strength of a cryptographic key that is used to protect data (i.e., a data-protection key), given the manner in which the key was generated and handled prior to its use to protect the target data. This document,

SP 800-158, *Key Management: Obtaining a Targeted Security Strength*, involves a considerable amount of new research since it is an area that has not been fully addressed to date.

Key Management:
http://csrc.nist.gov/groups/ST/key_mgmt/

Contacts:

Ms. Elaine Barker
(301) 975-2911
ebarker@nist.gov

Mr. Quynh Dang
(301) 975-3610
qdang@nist.gov

Dr. Lily Chen
(301) 975-6974
llchen@nist.gov

Dr. Allen Roginsky
(301) 975-3603
roginsky@nist.gov

Digital Signatures

FIPS 186, *The Digital Signature Standard*, specifies three techniques for the generation and verification of digital signatures that can be used for the protection of data: the Digital Signature Algorithm (DSA), the Elliptic Curve Digital Signature Algorithm (ECDSA), and the Rivest-Shamir Adelman (RSA) algorithm. A digital signature is represented in a computer as a string of bits and is computed using a set of rules and a set of parameters that allow the identity of the signatory and the integrity of the data to be verified.

FIPS 186, first published in 1994, has been revised several times since then, and in April 2012, a list of changes to the current version of the standard, FIPS 186-3, was provided for public comment. The proposed changes included a reduction of restrictions on the use of random number generators and the retention and use of prime number generation seeds, and further aligns the FIPS with Public Key Cryptography Standard (PKCS) #1. The comments received have been incorporated into the draft of a new version of the standard, which will be proposed to the Secretary of Commerce for adoption as FIPS 186-4 in FY2013. FIPS 186-3 and the proposed changes are provided at http://csrc.nist.gov/publications/PubsFIPS.html

---

Contacts:

Ms. Elaine Barker
(301) 975-2911
ebarker@nist.gov

Dr. Allen Roginsky
(301) 975-3603
roginsky@nist.gov

Block Cipher Modes of Operation

The engine for many of the techniques in NIST's cryptographic toolkit is a block cipher algorithm, such as the Advanced Encryption Standard (AES) algorithm or the Triple Data Encryption Algorithm (TDEA). A block cipher transforms data of a fixed length, called the block size, into seemingly random data of the same length. There are many cryptographic methods that feature block ciphers to achieve an information service, such as confidentiality or authentication. Such a method is called a block cipher mode of operation, or, simply, a mode.

NIST has formally approved modes in the area of "key wrapping", i.e., the protection of the confidentiality and integrity of cryptographic keys. NIST Special Publication (SP) 800-38F, *Recommendation for Block Cipher Modes of Operation: Methods for Key Wrapping*, describes existing methods that are approved for key wrapping, and also specifies three deterministic authenticated-encryption modes: the AES Key Wrap (KW) mode, the AES Key Wrap with Padding (KWP) mode, and one TDEA mode, called TKW.

NIST is also developing a set of modes in the area of "format preserving encryption" (FPE). A format is a sequence of decimal digits, such as a credit card number or a social security number; formats can also be defined for other sets of characters besides decimal digits. FPE is expected to be very useful because, in order to retrofit encryption to existing applications, it is sometimes necessary for the encrypted data to have the same format as the original data.

NIST intends to approve three schemes of the FFX framework for FPE that were submitted for consideration in recent years: FFX-base, VAES3, and BPS. (The submission documents are available at http://csrc.nist.gov/groups/ST/toolkit/BCM/modes_development.html). A draft SP to specify and approve these methods is expected to be ready for public comment in early 2013.

Contact:

Dr. Morris Dworkin
(301) 975-2354
morris.dworkin@nist.gov

→ Crypto Research

Post Quantum Cryptography

In recent years, there has been a substantial amount of research on quantum computers – machines that exploit quantum mechanical phenomena to solve problems that would be intractable for conventional computers. An early breakthrough in this area was Shor's algorithm, which demonstrated that quantum computers could efficiently factor integers and compute discrete logarithms. These two problems play an essential role in cryptography: they are believed to be hard for classical computers, and they are the basis for nearly all of the public-key cryptosystems that are in widespread use today. If large-scale quantum computers are ever built, they will be able to break the existing public-key infrastructure.

The threat posed by quantum computers appears to be serious, but not immediate. While there has been dramatic progress in experimental quantum physics, the construction of large-scale quantum computers still seems to be many years away. Moreover, the discovery of Shor's algorithm has also motivated researchers to propose so-called "post-quantum" cryptosystems – public-key cryptosystems that would be secure against quantum computers. It is hoped that these cryptosystems will allow us to maintain the public-key infrastructure in a world with quantum computers. For these reasons, NIST has started a project on post-quantum cryptography, with a view to possible future standards.

The primary focus of this project is to identify candidate quantum-resistant systems, based on algebraic codes, lattices, multivariate systems of equations, cryptographic hash functions, or any other construct that may be secure against both quantum and classical computers, as well as the impact that such post-quantum algorithms will have on current protocols and security infrastructures. The project endeavors to establish the viability of algorithms in these areas, the security of which have yet to be explained well even in the classical model, and further, to verify the claims of quantum-resistance as quantum complexity theory matures. In the event that no candidate algorithm survives this examination, NIST intends to establish computer security architectures that are not dependent upon the classical public-key cryptographic algorithms, such as the Rivest-Shamir-Adleman (RSA) algorithm or the Elliptic Curve Digital Signature Algorithm (ECDSA) algorithm.

In FY2012, NIST researchers Stephen Jordan, Yi-Kai Liu, Ray Perlner, and Daniel Smith-Tone internally presented preliminary status reports in the areas of quantum computation, coding-based cryptography, lattice-based cryptography, and multivariate cryptography, which included detailed surveys of the respective fields, as well as security overviews and specific results. These reports were further supplemented with a presentation from William Whyte and John Schanck from NTRU Cryptosystems on April 25, 2012, discussing the specific countermeasures being deployed in the wake of a serious attack on NTRUSign. NIST also engaged the international cryptographic community with presentations and publications by NIST researchers. At the very end of FY2011, on September 23, 2011, Stephen Jordan presented "Complexity Implications of Quantum Field Theory," at the Schloss Dagstuhl Workshop on Quantum Cryptanalysis, discussing evidence that more modern quantum field theories may not give rise to greater computational power than the standard quantum circuit model. In the first quarter of FY2012, Daniel Smith-Tone published the paper, "On the Differential Security of Multivariate Cryptosystems," at the Fourth International Conference on Post-Quantum Cryptography, suggesting a new security metric for multivariate cryptography. Daniel Smith-Tone also published, "The TriTon Transformation," discussing risky design philosophies in multivariate cryptography at the Third Workshop on Mathematical Cryptology on July 9, 2012. On September 28, 2012, at the Quantum Information Science workshop at the NIST-University of Maryland (UMD) Joint Quantum Institute, Yi-Kai Liu gave a talk on "Applications of Quantum Information in Machine Learning and Cryptography," which discussed the role played by quantum information in security proofs for lattice-based cryptosystems.

In FY2013, NIST will continue to explore the security capacity of purported quantum-resistant technologies with the ultimate goal of uncovering the fundamental mechanisms necessary for efficient, trustworthy, and cost-effective information assurance in the post-quantum market. Upon the successful completion of this phase of the project, NIST will be prepared for possible standardization.

Contacts:

Dr. Daniel Smith-Tone
(502) 852-6010
daniel.smith@nist.gov

Dr. Lily Chen
(301) 975-6974
lily.chen@nist.gov

Mr. Ray Perlner
(301) 975-3357
ray.perlner@nist.gov

Dr. Dustin Moody
(301) 975-8136
dustin.moody@nist.gov

Dr. Yi-Kai Liu
(301) 975-6499
yi-kai.liu@nist.gov

NIST Beacon – A Prototype Implementation of a Randomness Beacon

NIST is developing a Secure Randomness Beacon that will broadcast full-entropy bit-strings. The Beacon could be used in many applications that require a secure random value (e.g., for privacy-enhanced cryptography, multiparty contract bidding, and tamper-proof voting), but is not intended for generating values that must remain secret, such as cryptographic keys. The Beacon will be designed to provide unpredictability, autonomy, and consistency. Unpredictability means that users cannot algorithmically predict bits before they are made available by the source. Autonomy means that the source is resistant to attempts by outside parties to alter the distribution of the random bits. Consistency means that a set of users can access the source in such a way that they are confident that they all receive the same random string.

Strings of bits produced by the Beacon will be posted in blocks of 512 bits every few seconds, with the number of seconds being an adjustable parameter that can vary from one second to a few minutes. Each such value will be provided as an output packet that is sequence-numbered, time-stamped, and signed, and includes the hash of the previous value, in order to

chain the sequence of values together and prevent an undetected change of an output package, even by the source. Each packet will be stored for subsequent online access.

In pursuit of this goal, a prototype implementation of a public source of randomness is being developed that is conformant to SP 800-90A, *Recommendation for Random Number Generation Using Deterministic Random Bit Generators*. The Beacon's engine uses multiple sources of entropy and leverages recently developed tests to validate an entropy source.

During FY2012, NIST continued working on implementing and enhancing the NIST Secure Randomness Beacon. CSD also initiated collaboration with NIST's Physical Measurement Laboratory (PML) to enhance the input source of entropy for the NIST Randomness Beacon by integrating at least one quantum source, also referred to as a "truly random number." The joint project, Quantum Randomness as a Secure Resource, received ITL's Innovations in Measurement Science (IMS) award.

The ITL and PML collaborative work toward the integration of at least one quantum-secure source of entropy for the NIST Randomness Beacon will continue in FY2013. A quantum source will be a sequence of truly random numbers that is guaranteed by the laws of physics to be unknowable in advance of its generation, and uncorrelated with anything in the universe. With such a quantum source, the NIST Randomness Beacon can be used for a variety of security and privacy applications that could lead to unprecedented levels of network security for confidential digital applications, thus setting the foundation for a trusted common standard.

http://www.nist.gov/itl/csd/ct/nist_beacon.cfm

http://www.nist.gov/pml/div684/random_numbers_bell_test.cfm

Contacts:

Dr. Michaela Iorga
(301) 975-8431
michaela.iorga@nist.gov

Dr. Rene Peralta
(301) 975-8702
rene.peralta@nist.gov

Pairing-Based Cryptography

Recently, what are known as "pairings" on elliptic curves have been a very active area of research in cryptography. A pairing is a function that maps a pair of points on an elliptic curve into a finite field. Their unique properties have enabled many new cryptographic protocols that had not previously been feasible.

In particular, identity-based encryption (IBE) is a pairing-based scheme that has received considerable attention. IBE uses some form of a person (or entity's) identification to generate a public key. This could be an email address, for instance. An IBE scheme allows a sender to encrypt a message without needing a receiver's public key to have been certified and distributed for subsequent use. Such a scenario is quite useful if the pre-distribution of public keys is impractical. Besides IBE, there are a number of other applications of pairing-based cryptography. These include many other identity-based cryptosystems (including signature schemes), key establishment schemes, functional and attribute-based encryption, and privacy-enhancing techniques, such as the use of anonymous credentials.

In 2008, NIST held a workshop on pairing-based cryptography. While the workshop showed that there was interest in pairing-based schemes, a common understanding was that further study was needed before NIST approved any such schemes. Starting in 2011, members of the Cryptography Technology Group (CTG) have conducted an extensive study on pairing-based cryptographic schemes. This included topics such as: the construction of pairing-friendly elliptic curves, a survey of pairing-based cryptographic schemes, implementation efficiency with respect to the required security, standard activities involving pairing-based schemes, use cases, and practical implications. This work was summarized in a technical report, presented in the first quarter of 2012. Throughout 2012, project members have been identifying use cases for pairing-based cryptography. At the NIST Cryptography for Emerging Technologies and Applications (CETA) Workshop in November 2011, there was a public call for feedback on potential use cases.

Pairing operations appear to be important tools for various cryptographic schemes used in cloud computing and privacy-enhancing environments. Besides IBE, other demanding applications have also motivated the continuation of this study. Short signatures and broadcast encryption are examples of such applications.

Contacts:

Dr. Dustin Moody
(301) 975- 8136
dustin.moody@nist.gov

Dr. Lily Chen
(301) 975-6974
lily.chen@nist.gov

Privacy-Enhancing Cryptography Project

Modern cryptography provides powerful tools for protecting private information, but current standards are often blunt instruments for privacy protection. There are many ways CSD can develop and standardize new methods to use cryptography that enhance privacy. For example, public-key certificates used for authentication often reveal more personally identifiable information about the certificate holder than is required for a given application.

What is often at issue in accessing data or resources is not the identity of the customer, but whether the customer is a member of an eligible group. Methods that allow a user to selectively reveal and prove only a specific property (such as that the user is at least 21 years old, has a particular place of residence, or citizenship) are approaching commercial practicality. Other techniques, such as those that will eventually allow us to search encrypted databases, are still in the research stage. However, these techniques are sufficiently advanced that it behooves us to take stock of the state of the art at this point. Still other techniques, such as those that allow us to hold sealed-bid auctions without ever opening the bids, are known to be practical, yet have received little attention by those that might benefit from them. Such applications fall within the scope of what are known as secure multiparty computations.

In FY2012, NIST held a workshop on Privacy-Enhancing Cryptographic Techniques to explore processes, procedures, and potential applications that could benefit from the ability to operate on encrypted data without decrypting it (see http://www.nist.gov/itl/csd/ct/pec-workshop.cfm). Participants at the workshop included scientists, privacy advocates, and policy experts. Having planted the seeds for cooperation among these different groups, CSD will continue to pursue this objective in FY2013.

Another major activity for this project was in support of the National Strategy for Trusted Identities in Cyberspace (NSTIC) (see http://www.nist.gov/nstic). Different cryptographic techniques that may be important for this initiative are being continuously evaluated.

Contact:
Dr. Rene Peralta
(301) 975-8702
peralta@nist.gov

Cryptography for Constrained Environments

Pervasive computing is the emerging area in which many highly constrained devices are interconnected, typically communicating wirelessly with one another, and working in concert to accomplish some task. These systems can be found in a wide variety of fields. Sample application areas include: sensor networks, medical devices, distributed control systems, and the Smart Grid. Security can be very important in all of these areas. For example, an unauthorized party should not be able to take control of an insulin pump or the brakes on a car. There are also privacy concerns, particularly in the area of Health IT.

Because the majority of the current cryptographic algorithms were designed for desktop/server environments, many of these algorithms do not fit into the constrained resources currently available. If current algorithms can be made to fit into the limited resources of constrained environments, their performance is typically not acceptable. A particular problem is the use of asymmetric (public key) algorithms. These algorithms tend to be much more computational and resource-intensive than can be easily accommodated in such constrained environments.

As a result, NIST is currently focusing on studying the

NIST-approved symmetric algorithms in constrained environments. Symmetric algorithms can be used to perform encryption for confidentiality, as well as to generate message authentication codes (MAC) for message authentication. NIST has implemented the current 256-bit version of the Secure Hash Algorithm (SHA-256) to provide a Hash-based Message Authentication Code (HMAC) for authentication. Additionally, NIST has implemented the Advanced Encryption Standard (AES) to provide confidentiality, as well as the cipher-based message authentication code (CMAC) mode for authentication. An outline of these plans was provided at the Workshop on Cryptography for Emerging Technologies and Applications hosted by NIST in November 2011.

During the next year, NIST will investigate algorithms other than those currently in its Cryptographic Toolbox to find algorithms that are optimized for operating in constrained environments. NIST will analyze the resource requirements and performance characteristics of these algorithms, and utilize these block ciphers as building blocks to perform other cryptographic functions beyond encryption.

Contact:
Mr. Lawrence Bassham
(301) 975-3292
lbassham@nist.gov

New Research Areas in Cryptographic Techniques for Emerging Applications

In FY2012, NIST explored a few new research areas in cryptographic techniques for emerging applications. In particular, the research focused on stream ciphers, secure group communications, group signatures, and circuit complexity.

(1) Stream Ciphers

Currently, the use of AES in the Output Feedback Mode (OFB) mode and the counter (CTR) mode are approved by NIST as block cipher-based stream ciphers. However, classical stream ciphers have performance advantages for software implementations that satisfy high-throughput requirements, or for hardware implementations with constrained resources. The security and performance of some of the well-understood stream ciphers will be studied during FY2013, with a focus on stream ciphers designed for constrained environments.

(2) *Secure Group Communications*

Secure group communications has been shown to be important in public-safety networks, smart grids, and sensor networks. The existing schemes proposed in the research literature, such as multicast encryption schemes and group key-distribution schemes, have been considered as general solutions, but are less scalable for practical applications. In FY2012, CSD looked into existing results and explored different application scenarios. The requirements and the restrictions were also discussed. In FY2013, NIST will pursue well-tailored solutions for secure group communications.

(3) *Group Signatures*

Group signatures have been investigated for more than two decades. In general, a group signature scheme allows a group member to generate a signature on behalf of the group without revealing information about the specific signer. Numerous schemes have been proposed and analyzed in the research literature. Such an anonymity feature is useful for security applications in cloud computing. In FY2013, NIST will further explore the features and underlying mathematical structures for the existing schemes.

(4) *Circuit Complexity*

Any function can be described as a circuit with operations modulo 2. If the circuit contains only additions, then the function is linear. Nonlinearity, which is fundamental to cryptographic applications, can be achieved only by the use of multiplications. The standard description of the AES S-Box, which is the nonlinearity component for AES, is that it does inversion in the field of 256 elements. The field's standard measure of nonlinearity

of a function F is the Hamming distance of the spectrum of F to the closest linear spectrum. A different measure of nonlinearity is simply the number of multiplications necessary and sufficient to compute the function. This measure is called "multiplicative complexity." CSD's research in Boolean circuit optimization has yielded circuits with optimal or near-optimal multiplicative complexity for a large class of functions. The resulting circuits have large linear components. CSD developed new heuristics for reducing the number of gates in these components. The net result is a significant reduction in the size and/or depth of many circuits used in cryptography. These include a circuit of depth 16 and size 128 for the AES S-Box, as well as reduced size/depth circuits for high-speed cryptography in characteristic 2. Additionally, circuits with a small number of multiplications can be used to significantly improve the communication complexity of secure multiparty computations, as well as the size of non-interactive zero-knowledge proofs of knowledge.

Contacts:

Dr. Meltem Sonmez Turan
(301) 975-4391
meltem.turan@nist.gov

Dr. Rene Peralta
(301) 975-8702
rene.peralta@nist.gov

Dr. Lily Chen
(301) 975-6974
llchen@nist.gov

Dr. James Nechvatal
(301) 975-5048
james.nechvatal@nist.gov

Dr. Dustin Moody
(301) 975-8136
dustin.moody@nist.gov

Workshop on Cryptography for Emerging Technologies and Applications

NIST hosted a workshop on Cryptography for Emerging Technologies and Applications on November 7-8, 2011. The purpose of the workshop was to identify the cryptographic requirements for emerging technologies and applications.

The workshop provided an opportunity for the government, industry, research and academic communities to identify cryptographic challenges encountered in their development of emerging technologies and applications, and to learn about NIST's current cryptographic research activities, programs, and standards development.

In preparation for the workshop, NIST called for the submission of abstracts that highlight the cryptographic challenges identified during the research and development of emerging technologies and applications. Examples of emerging or evolving technology spaces include: sensor and building networks, mobile devices, smart objects/Internet of things, and cyber physical systems (CPSs). Examples of cryptographic requirements for emerging sectors include performance or resource issues, cryptographic services (such as anonymous or group signatures), or key management challenges. Abstract submitters were also encouraged to identify, through their submissions, other areas of cryptography for emerging technologies and applications.

NIST received twenty-eight abstracts. Out of these abstracts, eleven were selected for presentation during the workshop. In addition to these presentations, the workshop included two keynote talks, an invited talk, and six presentations by NIST. An open floor discussion concluded the workshop. The workshop agenda and its slide presentations are available at http://www.nist.gov/itl/csd/ct/ceta-2011-agenda.cfm.

Contacts:

Dr. Michaela Iorga
(301) 975-8431
michaela.iorga@nist.gov

Mr. Quynh Dang
(301) 975-3610
quynh.dang@nist.gov

Ms. Elaine Barker
(301) 975-2911
ebarker@nist.gov

→ Applied Cryptography

Development of Federal Information Processing Standard (FIPS) 140-3, Security Requirements for Cryptographic Modules

The FIPS 140, *Security Requirements for Cryptographic Modules*, standard defines the security requirements for cryptographic modules and is applicable to all federal agencies that use cryptography-based security systems to protect sensitive information in computer and telecommunication systems (including voice systems) as defined in Section 5131 of the Information Technology Management Reform Act of 1996, Public Law 104-106, and the Federal Information Security Management Act of 2002, Public Law 107-347. The standard must be used in designing and implementing cryptographic modules that federal departments and agencies operate or are operated for them under contract.

The current version of the standard is FIPS 140-2. A proposed revision, FIPS 140-3, will supersede FIPS 140-2, and is under development as a result of the reexamination and reaffirmation of FIPS 140-2. The draft revision of the standard adds new security requirements that are imposed on cryptographic modules to reflect the latest advances in technology and security, and to mirror other new or updated standards published by NIST in the areas of cryptography and key management. Additionally, software and firmware requirements are addressed in a new area dedicated to software and firmware security, while another new area specifying requirements to protect against non-invasive attacks is also provided.

Draft FIPS 140-3 provides four increasing qualitative levels of security that are intended to cover a wide range of potential applications and environments. The security requirements cover areas related to the secure design and implementation of a cryptographic module. These areas include cryptographic module specification; cryptographic module physical ports and logical interfaces; roles, authentication, and services; software security; operational environment; physical security; physical security - non-invasive attacks; sensitive security parameter management; self-tests; life-cycle assurance; and mitigation of other attacks. The standard provides users with a specification of security features that are required at each of four security levels, flexibility in choosing security requirements, a guide to ensuring that the cryptographic modules incorporate necessary security features, and the assurance that the modules are compliant with cryptography-based standards.

During FY2011, the majority of the resolutions to the public comments received on the second draft were incorporated in the draft FIPS and provided to NIST and CSE Canada for a final internal technical review. In response to comments received on the second draft, the following possible changes or additions to the previous draft document were proposed: a description of the assumed thread models for each security level; an insertion of missing definitions for terms and acronyms; changes to the Trusted Channel requirements; the removal of the Trusted Role; the inclusion of an identity-based authentication mechanism that would be allowed at Security Level (SL) 2; the addition of a self-initiated cryptographic output capability and remote control capability; the inclusion of additional integrity-technique requirements for the software components of a cryptographic module; a restructure of the annexes and enhancement of the requirements for the allowed operator-authentication mechanisms; an update of the list of the non-invasive attacks methods for the security functions; and an update of the requirements for the allowed modifiable operating environments.

During the process of addressing the public comments received on the second draft, NIST determined that additional feedback would be required from the public to resolve gaps and inconsistencies between the comments received for particular sections of the second draft of FIPS 140-3. As a result, NIST requested additional public comments in August 2012 on several clearly identified sections. More details about the project and a timetable can be found at: http://csrc.nist.gov/groups/ST/FIPS140_3/.

During FY2013, all received comments on the identified issues will be addressed and the final FIPS 140-3 document will be prepared for a final internal review and approval by the Secretary of Commerce.

http://csrc.nist.gov/groups/ST/FIPS140_3/

Contact:
Dr. Michaela Iorga
(301) 975-8431
michaela.iorga@nist.gov

Personal Identity Verification (PIV) Test Cards

Federal Information Processing Standard (FIPS) 201, *Personal Identity Verification (PIV) of Federal Employees and Contractors*, was published in February 2005 to satisfy policy directives specified in Homeland Security Presidential Directive-12 (HSPD-12). The majority of federal workers now have Personal Identity Verification (PIV) cards; however, the PIV card has not yet been embraced as a mechanism for logical access to IT resources. The unavailability of test PIV cards has been identified as an impediment to deployment for this purpose.

To facilitate the development of applications and middleware that support the PIV card, CSD developed a reference set of smart cards. Each set includes sixteen cards: nine valid cards and seven cards that contain invalid data. The valid cards differ in terms of the cryptographic algorithms used to sign the data objects, the types and sizes of the cardholder's key pairs, and in the presence or absence of optional data objects. The invalid cards include cards that are expired, cards that have certificates that have been revoked, and cards with data objects that have invalid signatures.

This set of test cards includes not only examples that are similar to cards that are currently issued today, but also examples of cards with features that are expected to appear in cards that will be issued in the future. For example, while the certificates and data objects on most, if not all, cards issued today are signed using RSA PKCS #1 v1.5, the set of test cards includes examples of certificates and data objects that are signed using each of the algorithms and key sizes approved for use with PIV cards, including the RSA Probabilistic Signature Scheme (RSASSA-PSS) and the Elliptic Curve Digital Signature Algorithm (ECDSA). Similarly, the infrastructure supporting the test cards provides examples of Certificate Revocation Lists (CRLs) and Online Certificate Status Protocol (OCSP) responses that are signed using each of these signature algorithms. The set of test cards also includes certificates with elliptic curve cryptography (ECC) subject public keys, in addition to RSA subject public keys, as is permitted by Table 3-1 of

SP 800-78-3, *Interfaces for Personal Identity Verification*. The set of test cards, collectively, also includes all of the mandatory and optional data objects listed in Section 3 of SP 800-73-3 Part 1, except for Cardholder Iris Images. Several of the cards include a Key History object, along with retired key management keys. The certificates that appear on the test cards, both the cardholders' certificates and the content signers' certificates, were issued from a simple two-level hierarchy.

The initial work of developing the test cards was performed during FY2011. In early FY2012, CSD created a few sets of test cards, which were distributed to organizations that had previously volunteered to serve as beta testers. During the beta-testing period, the test cards were used in a few different environments, and it was determined that no changes needed to be made to the specifications for the cards. In late FY2012, NIST began the production of the final sets of test cards, and the cards are now available as NIST Special Database 33 (http://csrc.nist.gov/groups/SNS/piv/testcards.html).

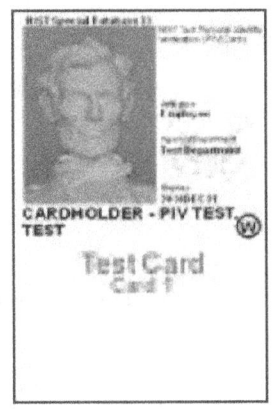

For further details on the PIV project, see the Personal Identity Verification (PIV) and FIPS 201 Revision Efforts section.

http://csrc.nist.gov/groups/SNS/piv/testcards.html

Contacts:

Dr. David Cooper
(301) 975-3194
david.cooper@nist.gov

Mr. William Polk
(301) 975-3348
william.polk@nist.gov

Authentication

To support the Office of Management and Budget (OMB) Memorandum M-04-04, E-Authentication Guidance for Federal Agencies, NIST developed SP 800-63, *Electronic Authentication Guideline*, and its subsequent revision, SP 800-63-1, was published at the end of last year. The OMB policy memorandum defines four levels of authentication in terms of assurance about the validity of an asserted identity. SP 800-63-1 gives technical requirements and examples of authentication technologies that meet the requirements by making individuals demonstrate possession and control of a secret for each of the four levels. In FY2013, NIST plans to develop a minor revision to the SP 800-63-1 identity-proofing requirements to better accommodate medical professionals.

In the course of developing the revision to the *Electronic Authentication Guideline*, NIST researchers have come across gaps that prevent large-scale adoption of secure authentication online. Two such areas are revocation in complex federated environments and biometric authentication in unattended scenarios.

With regard to revocation, a large number of Identity Management Systems (IDMSs) are being deployed worldwide that use different technologies for the population of their users. With the diverse set of technologies, and the unique business requirements for organizations to federate, there is no uniform approach to the federation process. Similarly, there is no uniform method to revoke credentials or their associated attribute(s) in a federated community. In the absence of a uniform revocation method, NIST researchers have been exploring and developing a model for revocation of credentials and attributes in a federated environment, with a particular focus on identifying missing requirements.

To address the use of biometrics in authentication for transactions online, NIST is considering high-level requirements for the use of biometrics in a multi-factor authentication framework, such as liveness detection (sometimes called biometric spoof detection), biometric template protection (for revoking and renewing biometric credentials), and web services standards for securely and uniformly handling biometric data online. NIST is leading multi-year projects and collaborating with the international research and standards communities in all three of these areas.

With respect to liveness detection, NIST is leading the development of the first standards activity on liveness detection and held a workshop to discuss a framework for measurement, interchangeable data, and testing in March 2012. On the topic of biometric template protection, a paper on "Criteria Towards Metrics for Benchmarking Template Protection Algorithms" was published at this year's International Conference on Biometrics, the result of a grant from NIST and collaboration with researchers in CSD. This work was followed up with a workshop in July 2012, co-hosted with Fraunhofer and the Biometrics Institute, at the Association Française de Normalisation (AFNOR) (co-located with a biometric standards meeting). Finally, NIST's continued efforts in the area of biometric web services have yielded NIST SP 500-288, *Specification for WS-Biometric Devices*, and a reference implementation of the Specification for Web Service Biometric Device (WS-BD) in both Java and .NET.

Contact:
Dr. Lily Chen
(301) 975-6974
llchen@nist.gov

Security in Wireless and Mobility Networks

Today, wireless networks often provide connections for end mobile devices using multiple and different radio technologies. In such a heterogeneous network, a mobile device may switch its connection to the network between different wireless technologies. Inter-technology handover has brought many challenges to existing security solutions, such as the delays caused by access authentication for each handover. New trust models for key management are also required. NIST has conducted intensive research in the security for media-independent handover and has actively participated in the IEEE 802 wireless standard activities. In FY2012, the mechanisms to provide services for proactive authentications and

key distributions for inter-technology handover have been standardized in Amendment 2 for the IEEE 802.21 Media Independent Handover Services. The mechanisms will enable secure media-independent handover in heterogeneous networks.

In FY2013, NIST will continue to conduct research on the security mechanisms for next-generation wireless networks and pursue security solutions for group management through participation in the standards activities of the IEEE 802.21d task group.

Contact:
Dr. Lily Chen
(301) 975-6974
llchen@nist.gov

Validation Programs

→ Cryptographic Programs and Laboratory Accreditation

The Cryptographic Algorithm Validation Program (CAVP) and the Cryptographic Module Validation Program (CMVP) were developed by NIST to support the needs of the user community for strong, independently tested, and commercially available cryptographic algorithms and modules. Through these programs, NIST works with private and governmental sectors and the cryptographic community to achieve security, interoperability, and assurance of correct implementation. The goal of these programs is to promote the use of validated algorithms, modules, and products and to provide federal agencies with a security metric to use in procuring cryptographic modules. The testing carried out by independent third-party laboratories accredited by the NIST National Voluntary Laboratory Accreditation Program (NVLAP) and the validations performed by the CMVP and CAVP programs provide this metric. Federal agencies, industry, and the public can choose cryptographic modules and/or products containing cryptographic modules from the CMVP Validated Modules List and have confidence in the claimed level of security and assurance of correct implementation.

Cryptographic algorithm and cryptographic module testing and validation are based on underlying published standards and guidance that are developed within the Computer Security Division (CSD) in collaboration with many other organizations. As federal agencies are required to use validated cryptographic modules for the protection of sensitive nonclassified information, the validated modules and the validated algorithms that the modules contain represent the culmination and delivery of the division's cryptography-based work to the end user.

The CAVP and the CMVP are separate, collaborative programs based on a partnership between NIST's CSD and the Communication Security Establishment Canada (CSEC). The programs provide federal agencies — in the United States and Canada — confidence that a validated cryptographic algorithm has been implemented correctly and that a validated cryptographic module meets a claimed level of security assurance. The CAVP and the CMVP validate algorithms and modules used in a wide variety of products, including secure Internet browsers, secure radios, smart cards, space-based communications, munitions, security tokens, storage devices, and products supporting Public Key Infrastructure (PKI) and electronic commerce. A module may be a standalone product, such as a virtual private network (VPN), smart card or toolkit, or one

module may be used in several products; as a result, a small number of modules may be incorporated within hundreds of products. Likewise, the CAVP validates cryptographic algorithms that may be integrated in one or more cryptographic modules.

The two validation programs (the CAVP and CMVP) provide documented methodologies for conformance testing through defined sets of security requirements. Security requirements for the CAVP are found in the individual validation system documents containing the validation test suites that are required to assure that the algorithm has been implemented correctly. The validation system documents are designed for each FIPS-approved and NIST-recommended cryptographic algorithm. Security requirements for the CMVP are found in FIPS 140-2, *Security Requirements for Cryptographic Modules*, and the associated test metrics and methods in Derived Test Requirements for FIPS 140-2. Annexes to FIPS 140-2 reference the underlying cryptographic algorithm standards or methods. Federal agencies are required to use modules that were validated as conforming to the provisions of FIPS 140-2. The CMVP developed Derived Test Requirements associated with FIPS 140-2 to define the security requirements and the test metrics and methods to ensure repeatability of tests and equivalency in results across the testing laboratories.

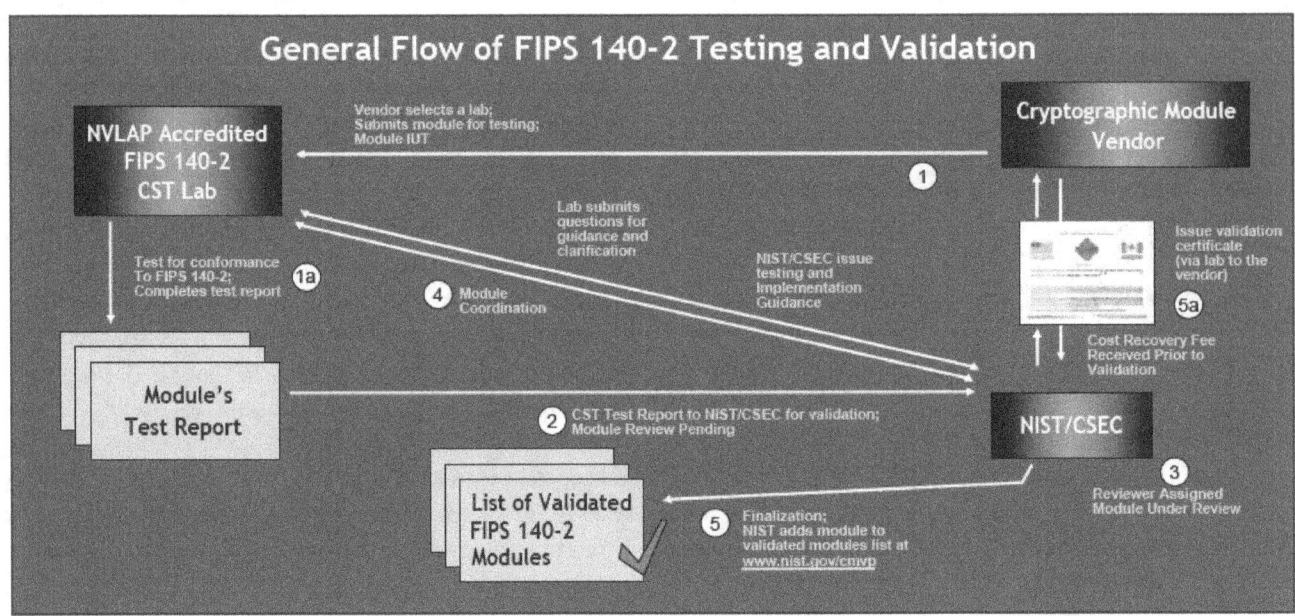

General Flow of FIPS 140-2 Testing and Validation

The CMVP reviews the cryptographic modules validation requests and, as a byproduct of the review, is attentive to emerging and/or changing technologies and the evolution of operating environments and complex systems during the module validation review activities. Likewise, the CAVP reviews the cryptographic algorithm validation requests submitted by the accredited laboratories. With these insights, the CAVP and CMVP can perform research and development of new test metrics and methods as they evolve. Based on this research, the CAVP and CMVP publish implementation guidance to assist vendors, testing laboratories, and the user community in the latest

quarterly from each of the testing laboratories, shows that 8 percent of the cryptographic algorithms and 61 percent of the cryptographic modules brought in for voluntary testing had security flaws that were corrected during testing. Without this program, the federal government would have had less than a 50 percent chance of buying correctly implemented cryptography. To date, over 1,850 cryptographic module validation certificates have been issued, representing over 4,275 modules that were validated by the CMVP. These modules have been developed by more than 400 domestic and international vendors.

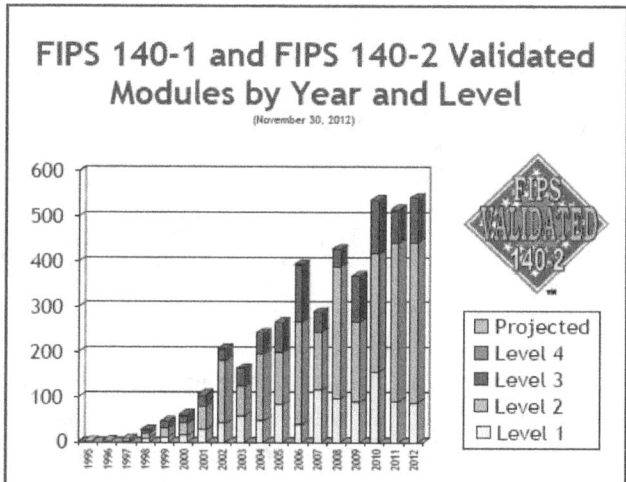

FIPS 140-1 and FIPS 140-2 Validated Modules by Year and Level
(November 30, 2012)

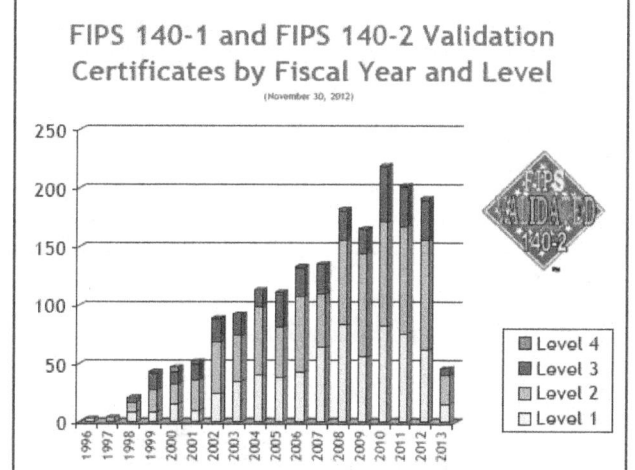

FIPS 140-1 and FIPS 140-2 Validation Certificates by Fiscal Year and Level
(November 30, 2012)

programmatic and technical guidance. This guidance provides clarity, consistency of interpretation, and insight for successful conformance testing, validation, and revalidation.

The unique position of the validation programs gives CMVP the opportunity to acquire insight during the validation review activities and results in practical, timely, and up-to-date guidance that is needed by the testing laboratories and vendors to move their modules and products out to the user community in a timely and cost-effective manner and with the assurance of third-party conformance testing. This knowledge and insight provide a foundation for future standards development.

The CAVP and the CMVP have stimulated improved quality and security assurance of cryptographic modules. The latest set of statistics, which are collected

The CAVP issued 2,225 algorithm validations and the CMVP issued 191 module validation certificates in FY2012. The number of algorithms and modules submitted for validation continues to grow, representing significant growth in the number of validated products expected to be available in the future.

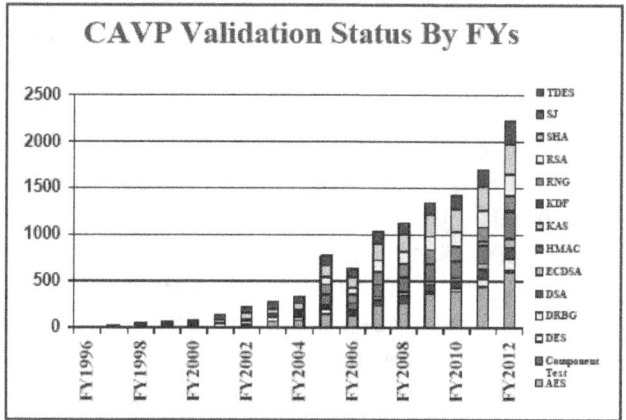

CAVP Validation Status By FYs

CAVP Validated Implementation Actual Numbers

Updated As Wednesday, October 03, 2012

FiscalYear	AES	Comp.	DES	DSA	DRBG	ECDSA	HMAC	KAS	KDF	RNG	RSA	SHA	SJ	TDES	Total
FY1996	0	0	2	0	0	0	0	0	0	0	0	0	0	0	2
FY1997	0	0	11	6	0	0	0	0	0	0	0	7	2	0	26
FY1998	0	0	27	9	0	0	0	0	0	0	0	6	0	0	42
FY1999	0	0	30	14	0	0	0	0	0	0	0	12	1	0	57
FY2000	0	0	29	7	0	0	0	0	0	0	0	12	1	28	77
FY2001	0	0	41	15	0	0	0	0	0	0	0	28	0	51	135
FY2002	30	0	44	21	0	0	0	0	0	0	0	59	6	58	218
FY2003	66	0	49	24	0	0	0	0	0	0	0	63	3	73	278
FY2004	82	0	41	17	0	0	0	0	0	28	22	77	0	70	337
FY2005	145	0	54	31	0	14	115	0	0	108	80	122	2	102	773
FY2006	131	0	3	33	0	19	87	0	0	91	63	120	1	83	631
FY2007	238	0	0	63	0	35	127	0	0	137	130	171	1	136	1038
FY2008	270	0	0	77	4	41	158	0	0	137	129	191	0	122	1129
FY2009	373	0	0	71	23	33	193	6	0	142	143	224	1	138	1347
FY2010	399	0	0	70	31	39	179	12	0	150	155	239	0	142	1416
FY2011	440	7	0	102	79	68	201	34	0	148	183	255	0	177	1694
FY2012	599	24	0	121	122	92	283	20	3	158	231	323	1	248	2225
Total	2773	31	331	681	259	341	1343	72	3	1099	1136	1909	19	1428	11425

http://csrc.nist.gov/groups/STM

Contacts:

CMVP Contact:
Mr. Randall J. Easter
(301) 975-4641
randall.easter@nist.gov

CAVP Contact:
Ms. Sharon Keller
(301) 975-2910
sharon.keller@nist.gov

→ Automated Security Testing and Test Suite Development

NIST's Cryptographic Technology Group (CTG) and Cryptographic Algorithm Validation Program (CAVP) work closely throughout the specification and guideline development process to develop guidance and specifications for cryptographic-based security that can be tested at the algorithm level to provide assurance that the algorithms are implemented correctly. The CTG is responsible for developing Federal Information Processing Standards (FIPS) and Special Publications (SPs), which contain guidance for cryptographic-based security. This includes the specification of approved cryptographic algorithms and the development of guidelines for using these cryptographic mechanisms to provide secure and reliable protection for federal information. NIST's Cryptographic Algorithm Validation Program (CAVP) is responsible for providing assurance that the algorithms are implemented correctly. The CAVP does this by designing and developing conformance testing for implementations of these algorithms.

The conformance tests consist of a suite of validation tests for each approved cryptographic algorithm. These tests exercise the mathematical formulas and the algorithmic requirements detailed in the algorithm to assure that the detailed specifications are implemented correctly and completely. If the implementer deviates from or excludes any part of these instructions or requirements, the validation test will fail, indicating that the algorithm implementation does not function properly or is incomplete.

CAVP developed validation tests are performed by accredited testing laboratories on a vendor's algorithm implementation using automated known-answer tests, which compare the result from a cryptographic operation with a specific input against the expected result. They provide a uniform way to assure that the cryptographic algorithm implementation adheres to the detailed specifications.

There are several types of validation tests, all designed to satisfy the testing requirements of the cryptographic algorithms and their specifications. These include, but are not limited to, Known-Answer Tests, Monte Carlo Tests, and Multi-Block Message Tests. The Known-Answer Tests are designed to test the conformance of the implementation under test (IUT) to the various specifications in the reference document. This involves testing the components of the algorithm to assure that they are implemented correctly. The Monte Carlo Test is designed to exercise the entire IUT. This test is designed to detect the presence of implementation flaws that are not detected with the controlled input of the Known-Answer Tests. The types of implementation flaws detected by this validation test include pointer problems, insufficient allocation of space, improper error handling, and incorrect behavior of the IUT. The Multi-Block Message Test (MMT) is designed to test the ability of the implementation to process multi-block messages, which require the chaining of information from one block to the next.

Automated security testing and test suite development are integral components of the CAVP. The CAVP encompasses validation testing for FIPS-approved and NIST-recommended cryptographic algorithms. Cryptographic algorithm validation is a prerequisite to the cryptographic module validation performed by the Cryptographic Module Validation Program (CMVP). The testing of cryptographic algorithm implementations is performed by independent third-party laboratories. The CAVP develops and maintains a Cryptographic Algorithm Validation System (CAVS) tool that automates the cryptographic algorithm validation testing.

During the last few years, the scope of requirements within the CTG's publications has expanded to include not only an algorithm's specification, but also how an algorithm should be used. Many of these requirements are outside the scope of the algorithm boundary and therefore cannot be tested at the algorithm level by the CAVP. Some of the requirements are routinely tested within NIST's Cryptographic Module Validation Program (CMVP), which validates cryptographic modules. Other requirements could potentially be tested by the CMVP, while others cannot; in the latter case, the fulfillment of the requirements is the responsibility of entities using, installing, or configuring applications or protocols that use the cryptographic algorithms. For example, depending on the design of a cryptographic module, it may not be possible for the module to determine whether or not a specific key is used for multiple purposes, a situation that is strongly discouraged.

The CAVS tool currently has algorithm validation testing for the following cryptographic algorithms:

Cryptographic Algorithm/Component	Special Publication or FIPS
Triple Data Encryption Standard (TDES)	SP 800-67, *Recommendation for the Triple Data Encryption Algorithm (TDEA) Block Cipher*, and SP 800-38A, *Recommendation for Block Cipher Modes of Operation-Methods and Techniques*
Advanced Encryption Standard (AES)	FIPS 197, *Advanced Encryption Standard*, and SP 800-38A
Digital Signature Standard (DSS)	FIPS 186-2, *Digital Signature Standard (DSS)*, with change notice 1, dated October 5, 2001
	FIPS 186-3, *Digital Signature Standard (DSS)*, dated June 2009
Elliptic Curve Digital Signature Algorithm (ECDSA)	FIPS 186-2, *Digital Signature Standard (DSS)*, with change notice 1, dated October 5, 2001 and ANSI X9.62
	FIPS 186-3, *Digital Signature Standard (DSS)*, dated June 2009 and ANSI X9.62
RSA algorithm	ANSI X9.31 and Public Key Cryptography Standards (PKCS) #1 v2.1: RSA Cryptography Standard-2002
	FIPS 186-3, *Digital Signature Standard (DSS)*, dated June 2009 and ANSI X9.31 and Public Key Cryptography Standards (PKCS) #1 v2.1: RSA Cryptography Standard-2002

Cryptographic Algorithm/Component	Special Publication or FIPS
Hashing algorithms SHA-1, SHA-224, SHA-256, SHA-384, SHA-512, SHA-512/224, SHA-512/256	FIPS 180-4, *Secure Hash Standard (SHS)*, dated March 2012
Random number generator (RNG) algorithms	FIPS 186-2 Appendix 3.1 and 3.2; ANSI X9.62 Appendix A.4
Deterministic Random Bit Generators (DRBG)	SP 800-90, *Recommendation for Random Number Generation Using Deterministic Random Bit Generators*
Keyed-Hash Message Authentication Code (HMAC)	FIPS 198, *The Keyed-Hash Message Authentication Code (HMAC)*
Counter with Cipher Block Chaining-Message Authentication Code (CCM) mode	SP 800-38C, *Recommendation for Block Cipher Modes of Operation: the CCM Mode for Authentication and Confidentiality*
Cipher-based Message Authentication Code (CMAC) Mode for Authentication	SP 800-38B, *Recommendation for Block Cipher Modes of Operation: The CMAC Mode for Authentication*
Galois/Counter Mode (GCM) GMAC Mode of Operation	SP 800-38D, *Recommendation for Block Cipher Modes of Operation: Galois/Counter Mode (GCM) and GMAC*, dated November 2007
XTS Mode of Operation	SP800-38E, *Recommendation for Block Cipher Modes of Operation: The XTS-AES Mode for Confidentiality on Block-Oriented Storage Devices*, dated January 2010
Key Agreement Schemes and Key Confirmation	SP 800-56A, *Recommendation for Pair-Wise Key Establishment Schemes Using Discrete Logarithm Cryptography*, dated March 2007
All of SP 800-56A except KDF	SP 800-56A All sections except Section 5.8 Key Derivation Functions for Key Agreement Schemes
SP 800-56A Section 5.7.1.2 ECC CDH function	SP 800-56A Section 5.7.1.2 Elliptic Curve Cryptography Cofactor Diffie-Hellman (ECC CDH) Primitive Testing
Key-Based Key Derivation functions (KBKDF)	SP800-108, *Recommendation for Key Derivation using Pseudorandom Functions*, dated October 2009

Cryptographic Algorithm/Component	Special Publication or FIPS
Application-Specific Key Derivation functions (ASKDF) (includes KDFs used by IKEv1, IKEv2, TLS, ANS X9.63-2001, SSH, SRTP, SNMP, and TPM	SP800-135 (Revision 1) *Recommendation for Existing Application-Specific key Derivation Functions*, dated December 2011
Component test – ECDSA Signature Generation of hash value (This component test verifies the signing of a hash-sized input. It does not verify the hashing of the original message to be signed.)	FIPS 186-3, *Digital Signature Standard (DSS)*, dated June 2009 and ANSI X9.62
Component test – RSA PKCS#1 1.5 Signature Generation of encoded message EM (This component test verifies the signing of an EM. It does not verify the formatting of the EM.)	FIPS 186-3, *Digital Signature Standard (DSS)*, dated June 2009 and Public Key Cryptography Standards (PKCS) #1 v2.1: RSA Cryptography Standard-2002
Component test – RSA PKCS#1 PSS Signature Generation of encoded message EM (This component test verifies the RSASP1 function.)	FIPS 186-3, *Digital Signature Standard (DSS)*, dated June 2009 and Public Key Cryptography Standards (PKCS) #1 v2.1: RSA Cryptography Standard-2002

In FY2013, the CAVP expects to augment the CAVS tool to provide, at a minimum, algorithm validation testing for:

* SP 800-56C, *Recommendation for Key Derivation through Extraction-then-Expansion*, dated November 2011;

* SP 800-132, *Recommendation for Password-Based Key Derivation Part 1: Storage Applications*, dated December 2010;

* Addition of SHA 512/224 and SHA 512/256 to other algorithms that utilize the hashing function, and

* SP 800-56A Revision 1, *Recommendation for Pair-Wise Key Establishment Schemes Using Discrete Logarithm Cryptography*, dated March 2007.

http://csrc.nist.gov/groups/STM/cavp

Contacts:

Ms. Sharon Keller
(301) 975-2910
sharon.keller@nist.gov

Ms. Elaine Barker
(301) 975-2911
elaine.barker@nist.gov

→ ISO Standardization of Security Requirements for Cryptographic Modules

CSD has contributed to the activities of the International Organization for Standardization/International Electrotechnical Commission (ISO/IEC), which issued ISO/IEC 19790, *Security Requirements for Cryptographic Modules*, on March 1, 2006, and ISO/IEC 24759, *Test Requirements for Cryptographic Modules*, on July 1, 2008. These efforts bring consistent testing of cryptographic modules to the global community.

ISO/IEC JTC 1/SC 27 WG 3 has progressed on the revision of ISO/IEC 19790 and the revision of ISO/IEC 24759 for which Randall J. Easter of CSD is the editor. The revision of 19790 was completed, and it was published August 15, 2012. ISO/IEC 19790:2012 was also adopted by the American National Standards Institute (ANSI). The first Committee Draft (CD) of 24759 was completed in July 2012 and circulated for national body comment. It is expected that the revision of 24759 will be published in FY2013.

Work is nearing completion on the Technical Report document, ISO/IEC 30104 "*Physical Security Attacks, Mitigation Techniques and Security Requirements,*" for which Randall J. Easter of CSD is the editor. A final draft of 30104 was completed in June 2012 and circulated for national body comment.

Work is progressing on a new standard document, ISO/IEC 17825 "*Testing methods for the mitigation of non-invasive attack classes against cryptographic,*" for which Randall J. Easter of CSD is the editor. The second working draft of 17825 was completed in June 2012 and circulated for national body comment.

National body comments for the above three documents will be addressed at the 45th SC 27 WG 3 meeting to be held in Rome, Italy, in October 2012.

A new work item was proposed (NWIP) at the 44th SC 27 WG 3 meeting, which was held in May 2012, to address "*Cryptographic algorithms and security mechanisms conformance testing.*" It is expected to be approved as a new work item at the 45th SC 27 WG 3 with Randall J. Easter of CSD appointed as editor.

http://csrc.nist.gov/groups/STM/cmvp/

Contact:
Mr. Randall J. Easter
(301) 975-4641
randall.easter@nist.gov

→ Security Content Automation Protocol (SCAP) Validation Program

The SCAP Validation Program performs conformance testing to ensure that products correctly implement SCAP as defined in SP 800-126, *The Technical Specification for the Security Content Automation Protocol (SCAP): SCAP Version 1.2*. Conformance testing is necessary because SCAP is a complex specification consisting of eleven individual specifications that work together to meet various use cases. A single error in product implementation could result in undetected vulnerabilities or policy noncompliance within agency and industry networks.

The SCAP Validation Program was created by request of the Office of Management and Budget (OMB) to support the Federal Desktop Core Configuration (FDCC) and United States Government Configuration Baseline (USGCB). The program coordinates its work with the NIST National Voluntary Laboratory Accreditation Program (NVLAP) to set up independent conformance testing laboratories that conduct the testing based on draft NISTIR 7511 Revision 3, *Security Content Automation Protocol (SCAP) Version 1.2 Validation Program Test Requirements*. When testing is completed, the laboratory submits a test report to CSD for review and approval. SCAP validation testing has been designed to be inexpensive, yet effective. The SCAP conformance tests are either easily human-verifiable or automated through NIST-provided reference tools. To date, the program has 9 accredited independent laboratories and has validated 50 products from 32 different vendors.

The SCAP Validation Program will expand in FY2013 to provide enhanced testing support, and will evolve to include new technologies as SCAP matures. Expansion plans include support for United States Government Configuration Baseline (USGCB) releases, public SCAP validation test content, and expanded automated testing capabilities.

http://scap.nist.gov/validation/

Contact:
Ms. Melanie Cook
(301) 975-5259
melanie.cook@nist.gov

→ Personal Identity Verification (PIV) and FIPS 201 Revision Efforts

In response to Homeland Security Presidential Directive-12 (HSPD-12), *Policy for a Common Identification Standard for Federal Employees and Contractors*, Federal Information Processing Standard (FIPS) 201, *Personal Identity Verification (PIV) of Federal Employees and Contractors*, was developed and was approved by the Secretary of Commerce in February 2005. HSPD-12 calls for the creation of a new identity credential for federal employees and contractors. FIPS 201 is the technical specification of both the PIV identity credential and the PIV system that produces, manages, and uses the credential. This work is done in collaboration with the Cryptographic Technology Group.

CSD activities in FY2012 directly supported the revision and maintenance of the FIPS 201 standard. CSD performed the following activities during FY2012 to revise the standard:

* Drafted and published a second public-comment draft of FIPS 201-2 on July 9, 2012. Revised draft FIPS 201-2 reflects the disposition of more than 1,000 comments received from over 40 organizations on the first public-comment draft of FIPS 201-2. NIST coordinated with the Office of Management and Budget (OMB) and other U.S. government (USG) stakeholders before incorporating changes in the revised draft FIPS 201-2;

* Organized and facilitated a workshop to discuss the contents of revised draft FIPS 201-2. NIST held a one-day workshop on July 25, 2012, to discuss contents of revised draft FIPS 201-2. The workshop was another mechanism to reach out to the PIV community, to interact with implementers and vendors, to clarify and explain changes in revised draft FIPS 201-2 as a result of comment dispositions, and to encourage the PIV community to provide formal comments to NIST; and

* Processed and analyzed comments received on revised draft FIPS 201-2. NIST started to review and process more than 500 comments received from over 30 organizations.

In FY2013, CSD will be focusing on completing the revision of draft FIPS 201-2 and updating the relevant Special Publications (SP) associated with FIPS 201-2. In addition to updating the relevant publications, CSD will also develop two new SPs: SP 800-156, *Representation of PIV Chain-of-Trust for Import and Export*, and SP 800-157, *Guidelines for Personal Identity Verification (PIV) Derived Credentials*. CSD will also continue to provide technical and strategic inputs to the PIV-related initiatives.

http://csrc.nist.gov/groups/SNS/piv

Contacts:
Ms. Hildegard Ferraiolo   Dr. David Cooper
(301) 975-6972   (301) 975-3194
hildegard.ferraiolo@nist.gov   david.cooper@nist.gov

→ NIST Personal Identity Verification Program (NPIVP)

The objective of the NIST Personal Identity Verification Program (NPIVP) is to validate PIV components for conformance to specifications in FIPS 201 and its companion documents. The two PIV components that come under the scope of NPIVP are PIV Smart Card Application and PIV Middleware. All of the tests under

NPIVP are handled by third-party laboratories that are accredited as Cryptographic and Security Testing (CST) Laboratories by the NIST National Voluntary Laboratory Accreditation Program (NVLAP) and are called accredited NPIVP test facilities. As of September 2011, there are ten such facilities.

In prior years, CSD published Special Publication (SP) 800-85A, *PIV Card Application and Middleware Interface Test Guidelines*, to facilitate development of PIV Smart Card Application and PIV Middleware that conform to interface specifications in SP 800-73, *Interfaces for Personal Identity Verification*. CSD also developed an integrated toolkit called "PIV Interface Test Runner" for conducting tests on both PIV Card Application and PIV Middleware products, and provided the toolkit to accredited NPIVP test facilities.

Throughout FY2012, the versions of the documents that were used as the basis for NPIVP validation are the following:

* *Card and Middleware Interface Specification* - SP 800-73-3; and
* *Card Application and Middleware Conformance Tests* - SP 800-85A-2.

In FY2012, six new PIV card application products were validated for conformance to SP 800-73-3, and were issued certificates, bringing the total number of NPIVP-validated PIV Card application products to 35. Two PIV Middleware products were validated for conformance to SP 800-73-3 and were issued certificates, bringing the total number of NPIVP-validated PIV Middleware products to 19.

In addition, NPIVP completed all planning tasks relating to update of all FIPS 201 companion documents as well as the PIV Test Runner toolkit consequent on the expected publication of FIPS 201-2.

http://csrc.nist.gov/groups/SNS/piv/npivp

Contacts:

Dr. Ramaswamy Chandramouli Ms. Hildegard Ferraiolo
(301) 975-5013   (301) 975-6972
mouli@nist.gov   hildegard.ferraiolo@nist.gov

Research in Emerging Technologies

→ Cloud Computing and Virtualization

Cloud computing offers the possibility to increase the efficiency of IT services, to decreased cost in terms of capital expenses (CAPEX) and operational expenses (OPEX), and to leverage leading-edge technologies to meet the information processing needs of the United States government (USG). However, the change in control dynamics poses new security challenges for the cloud computing adopters.

To accelerate the federal government's secure adoption of cloud computing, NIST assumed the leading role in developing standards and guidelines in close consultation and collaboration with standards bodies, the private sector, and other stakeholders. NIST's long-term goal is to provide thought leadership and guidance around the cloud computing paradigm to catalyze its use within industry and government. The NIST area of focus is technology, and specifically, interoperability, portability, and security requirements, standards, and guidance.

NIST Cloud Computing Program Support
NIST Cloud Computing Program strategically prioritizes NIST tactical projects that support USG agencies in their missions of secure and effective cloud computing adoption.

During FY2012, the NIST Cloud Computing team continued to promote the development of national and international standards and specifications that support USG's effective and secure use of cloud computing and to provide technical guidance to USG agencies for a secure and effective cloud computing adoption.

The CSD members of the NIST cloud computing team contributed to the research and development efforts of several public working groups which focused their activities on:

* Surveying the existing standards landscape for security, portability, and interoperability standards/models/studies/etc. relevant to cloud computing, determining standards gaps, and identifying standardization priorities, to develop a NIST Cloud Computing

Standards Roadmap that can be incorporated into the USG Cloud Computing Technology Roadmap (the *Standards Roadmap* Working Group);

* Developing a USG Cloud Computing Technology Roadmap that defines and prioritizes USG requirements for interoperability, portability, and security for effective cloud computing adoption (the *Reference Architecture and Taxonomy* Working Group);

* Developing a NIST Cloud Computing Security Reference Architecture - a framework and methodology for the secure adoption of cloud computing, which supplements the NIST Reference Architecture (*Security* Working Group);

* Supporting the cloud computing groups under the Federal CIO Council, providing technical advice to the Cloud Computing Executive Steering Committee, to the Cloud Computing Advisory Council, to the Information Security and Identity Management, and the Web 2.0 working group (the *Federal Cloud Computing Standards and Technology* Working Group); and

* Formulating a strategy for facilitating the development of high-quality cloud computing standards and describing a process for formulating cloud computing use cases and for judging the extent to which cloud system interfaces can satisfy them (the *Standards Acceleration to Jumpstart Adoption of Cloud Computing* (SAJACC) Working Group).

During FY2012, the NIST cloud computing team supported the fourth Cloud Computing Forum held in June 2012, at the Department of Commerce in Washington, D.C.

The cloud computing team contributed to the NIST Standards Working Group efforts of developing the draft Special Publication (SP) 500-292, *NIST Reference Architecture*, and draft SP 500-293, *U.S. Government Cloud Computing Technology Roadmap (volumes 1, 2, and 3)*.

The CSD cloud computing team also led the NIST Security Working Group's task of developing the *NIST Cloud Computing - Security Reference Architecture* working document, and contributed to the development of the white paper *"Challenging Security Requirements for the USG Cloud Computing Adoption."*

In FY2012, the leadership of the SAJACC Public Working Group, previously provided by CSD, was transitioned to other members of the Working Group, while the governance was kept with the NIST Cloud Computing Program.

In FY2012, the NIST cloud computing team also presented the results of cloud computing research and development, introduced the standards and specifications under development, and provided status of the NIST Cloud Computing Program in a variety of conferences and workshops.

Leveraging Access Control for Cloud Computing

In 2012, CSD continued the extensive research and development of a virtualization-based, enterprise-wide controlled delivery of data services for advanced cloud computing through Access Control (AC). Data services (DSs) are capabilities that enable the reading, manipulation, computation, presentation, management, and sharing of data. Typical DSs include applications such as email, workflow management, enterprise calendar, and records management, as well as system-level features, such as file, access control, and identity management. Although access control currently plays an important role in securing DSs, if properly envisaged and designed, AC can serve a more vital role in computing than one might expect. The Policy Machine (PM), a framework for AC developed at NIST, was designed with this goal in mind. The PM has evolved beyond just a concept to a prototype implementation and is now (FY2012 & FY2013), being implemented in a virtualized environment providing cloud-like features.

To appreciate the PM's advantages in computing, it is important to recognize the methods in which DSs are delivered today. Each DS runs in an Operating Environment (OE) and an OE can be of many types (e.g., operating systems, web services, middleware, and database and database applications), each implementing its own routines to enable the execution of DS-specific operations (e.g., read, send, and view) on their respective data types (e.g., files, messages, and fields).

This heterogeneity among OEs introduces a number of administrative and policy enforcement challenges and user inconveniences. Administrators must contend with a multitude of security domains when managing privileges, and ordinary users and administrators alike must authenticate to and establish sessions within different OEs in order to exercise legitimate DS capabilities. Even if properly coordinated across OEs, access control policies are not always globally enforced. An email application may, for example, distribute files to users regardless of an operating system's protection settings on those files. Also, while researchers, practitioners, and policy makers have specified a large variety of access control policies to address real-world security issues, only a relatively small subset of these policies can be enforced through off-the-shelf technology, and even a smaller subset can be enforced by any one OE.

It is the CSD Cloud Computing team's experience that the PM can provide an enterprise-wide OE that dramatically alleviates many of the administrative, policy enforcement, data interoperability, and usability issues that enterprises face today.

In particular, the cloud infrastructure is an OE in which the PM's functional components run in virtual machines. In this deployment, users and data objects can be provisioned, and DSs can be selected by the subscriber. DSs can be provided as SaaS or PaaS if they conform to the PM's API. CSD PM-motivated cloud differs from other types of clouds in the properties that it provides—users and objects are global, the framework is object type agnostic, DSs naturally interoperate, and AC policies are managed and enforced comprehensively across all DSs. CSD's cloud is also different in the degree of control that it offers to its subscribers. AC Policies can be imported from a library of predefined configurations, or can be configured from scratch by the subscriber, conferring PM the attributes of a Policy-as-a-Service (POLICYaaS) provider. POLICYaaS supports a wide range of policies including well-documented policies

such as Role-based Access Control, Discretionary Access Control, and Mandatory Access Control, as well as combinations of those policies. POLICYaaS can also accommodate separation of duty, conflict-of-interest, data tracking, and confinement policies, and should likely accommodate other unanticipated policies of the future.

The practical advantages of this PM-enabled cloud are many. Through a single authenticated session, users are offered capabilities of a variety of DSs to include office applications, file management, email, workflow, and records management. Data is naturally protected across DSs. Instead of deploying and managing different AC schemes for different DSs, select capabilities (of different DSs) are delivered to select users, under combinations of arbitrary, but mission-tailored forms of discretionary, mandatory, and history-based ACs. This interoperability property is not achieved through features or interfaces built into the DS, but rather through the OE that inherently provides a foundational basis for interoperability.

Virtualization Security & Leveraging Virtualization for Security

Virtualization is one of the foundational technologies that facilitate the use of a computing infrastructure for cloud computing services. At the core of a virtualized infrastructure is the virtualized host that provides abstraction of the hardware (i.e., CPU, memory, etc.) enabling multiple computing stacks (made up of O/S, Middleware and Applications) to be run on a single physical machine. The software that provides the abstraction features as well as the management capability to run multiple computing stacks (called Virtual Machines or VMs) is called the Hypervisor. The hypervisor also provides functions to define an entirely software-defined network inside a virtualized host (called a Virtual Network) for enabling communication among VMs running inside the virtualized host as well as to enable connectivity with the enterprise network outside the host. In addition, virtualization can also be implemented for the data storage infrastructure as well.

Taking into account the widespread scope for virtualization in the IT infrastructure, CSD has taken a comprehensive view of the security implications of this technology through a multifaceted approach that includes the following perspectives.

Perspective 1: Ensuring protection of the core virtualization module - the Hypervisor through a combination of architectural choices, configuration options, and operational practices.

Perspective 2: Ensuring protection of applications and O/S (called guest O/S) running inside virtual machines (VMs) through a combination of segmentation of the virtual network, virtual firewalls, Anti-Virus/Anti-Malware software, and IDS/IPS devices.

Perspective 3: Leveraging virtualization features to enhance security protection wherever possible.

In support of Perspective 1, in FY2012, NIST has developed a preliminary draft of NIST IR 7852, *Secure Management Practices for Protection of Hypervisors,* which outlines a set of recommendations for secure deployment of the complete hypervisor platform. Looking at security aspects from Perspective 2, one of the objectives was to identify the differences in security protection measures for VMs that are available to stakeholders between virtualized infrastructures deployed entirely for internal use from the ones that are used for offering cloud computing services. Towards this objective, a peer-reviewed conference paper titled "Security Control Variations between In-house and Cloud-based Virtualized Infrastructures" was written and presented at the 5th International Conference of Dependability.

In FY2013, CSD plans to obtain feedback and comments and publish the final version of NIST IR 7852. The core tasks in the area of virtualization for FY2013 will consist of security analysis and security recommendations based on all three perspectives described above, through the medium of conference papers and NIST publications, so as to promote secure adoption of this critical technology.

NIST Cloud Computing Program:
http://www.nist.gov/itl/cloud
http://collaborate.nist.gov/twiki-cloud-computing/
bin/view/CloudComputing/WebHome

Contacts:
Dr. Michaela Iorga
(301) 975-8431
michaela.iorga@nist.gov

Leveraging Access Control
for Cloud Computing:
Mr. David Ferraiolo
(301) 975-3046
david.ferraiolo@nist.gov

Virtualization Security &
Leveraging Virtualization for Security
Dr. Ramaswamy Chandramouli
(301) 975-5013
mouli@nist.gov

→ Mobile Device Security

Smart phones have become both ubiquitous and indispensable for consumers and business people alike. Although these devices are relatively small and inexpensive, they can be used not only for voice calls and simple text messages, but also for many functions once limited to laptop and desktop computers. Smart phones and tablet devices have specialized built-in hardware, such as photographic cameras, video cameras, accelerometers, Global Positioning System (GPS) receivers, and removable media readers. Furthermore, they employ a range of wireless interfaces, including infrared, Wireless Fidelity (Wi-Fi), Bluetooth, Near Field Communications (NFC), and one or more types of cellular interfaces that provide network connectivity across the globe. Although small in terms of form-factor, they can be used for sending and receiving email, browsing the web, online banking and commerce, social networking, storing and modifying documents, remotely accessing data, recording audio and video, and as navigation aids. Naturally, just as consumers and business people can realize productivity gains from these technologies, so can government agencies.

Like any new technology, smart phones present new capabilities, but also a number of new security challenges. Moreover, as the pace of the technology life cycles continues to increase, current Information

Assurance standards and processes must be updated and new technologies developed to transition from the use of specialized Government Off-The-Shelf (GOTS) products to Commercial Off-The-Shelf (COTS) products to allow government users to employ the latest and greatest technologies that consumers can use without sacrificing any privacy and security.

NIST is conducting research in new testing methodologies for smart phone software (apps) and is working with industry to bridge the security gaps present on today's smart phones. NIST has developed an online beta App Testing Portal for Android that examines app functionality with respect to agency security and privacy guidelines.

NIST will be publishing the following Special Publications (SP) in FY2013:

* SP 800-124, *Guidelines for Managing and Securing Mobile Devices in the Enterprise*;
* SP 800-164, *Guidelines on Hardware-Rooted Security in Mobile Devices*; and
* SP 800-163, *Guidelines for Testing and Vetting Mobile Applications*.

Contacts:

Dr. Steve Quirolgico
(301) 975-8426
stephen.quirolgico@nist.gov

Dr. Jeffrey Voas
(301) 975-6622
jeff.voas@nist.gov

Dr. Tom Karygiannis
301-975-4728
karygiannis@nist.gov

Strengthening Internet Security

➔ **USGv6: A Technical Infrastructure to Assist IPv6 Adoption**

Internet Protocol Version 6 (IPv6) is an updated version of the current Internet Protocol, IPv4. The primary motivations for the development of IPv6 were to increase the number of unique IP addresses and to handle the needs of new Internet applications and devices. In addition, IPv6 was designed with the following goals: increased ease of network management

and configuration; expandable IP headers; improved mobility and security; and quality of service controls. IPv6 has been, and continues to be, developed and defined by the Internet Engineering Task Force (IETF).

This year was a significant year for the deployment of IPv6 in the United States government. Office of Management and Budget's (OMB) Memo of September 10, 2010, entitled "Transition to IPv6," required all government agencies to "upgrade public/external facing servers and services (e.g., web, email, Domain Name System [DNS], Internet Service Provider [ISP] services, etc.) to operationally use native IPv6 by the end of FY2012." NIST has been working with the USGv6 Task Force and with individual government agencies to achieve this goal. NIST has developed an online monitor to demonstrate which high-level government domains have met this goal with respect to DNS services, email, web servers, and DNSSEC. OMB is using this monitor to measure USGv6 compliance with their latest milestone.

The NIST IPv6 Test Program, whose goal is to provide assurance on IPv6 conformance and interoperability of products, continued to operate. Additional tests were added, and the Supplier's Declaration of Conformity (SDOC), the vehicle used to enable vendors of IPv6 products to report the details of their products that have successfully executed the United States Government IPv6 (USGv6) tests, was improved.

In FY2013, NIST will continue to manage and evolve the USGv6 Test Program, and will update the NIST IPv6 Profile.

http://www.antd.nist.gov/usgv6

Contacts:

Ms. Sheila Frankel
(301) 975-3297
sheila.frankel@nist.gov

Mr. Douglas Montgomery
(301) 975-3630
dougm@nist.gov

→ Access Control and Privilege Management Research

With the advance of current computing technologies and the multifaceted environments the technologies are applied to, security issues such as situation awareness, trust management, privacy control for access control, and privilege management systems are becoming more complex. However, the research available on these topics is generally targeted to a specific system, is incomplete, makes assumptions, or is ambiguous regarding critical elements. Thus, practical and conceptual general guidance for these topics is needed.

In FY2012, CSD completed the development of the NIST Interagency Report (NISTIR) 7874, *Guidelines for Access Control System Evaluation Metrics*; researched unified enforcement mechanism of data services from Policy Machine (PM) for Enterprise Computing environment; enhanced the capabilities of the Access Control Policy Tool (ACPT); researched new access control rule verification method using diagraph algorithms; and started the development of draft NIST Special Publication (SP) 800-162, *Attribute-Based Access Control (ABAC) Definition and Considerations*, which provides information explaining the applications as well as formal models of ABAC.

In FY2013, CSD will continue the development of draft NIST SP 800-162; research unified enforcement mechanisms of data services from Policy Machine (PM) for Enterprise Computing environments; enhance the capabilities of the Access Control Policy Tool (ACPT); and research diagraph algorithms for policy rule composition of ABAC.

CSD expects that this project will:

* Promote (or accelerate) the adoption of community computing that utilizes the power of shared resources and common trust management schemes;

* Provide a standard evaluation metric in evaluating or comparing access control mechanisms for implementing access control applications;

* Increase security and safety of static (connected) distributed systems by applying the testing and verification tool for the access control policies; and

* Assist system architects, security administrators, and security managers whose expertise is related to access control or privilege policy in managing their systems, and in learning the limitations and practical approaches for their applications.

Contacts:

Dr. Vincent Hu
(301) 975-4975
vhu@nist.gov

Mr. David Ferraiolo
(301) 975-3046
david.ferraiolo@nist.gov

Mr. Rick Kuhn
(301) 975-3337
kuhn@nist.gov

→ Conformance Verification for Access Control Policies

Access control systems are among the most critical network security components. Faulty policies, misconfigurations, or flaws in software implementation can result in serious vulnerabilities. The specification of access control policies is often a challenging problem. Often a system's privacy and security are compromised due to the misconfiguration of access control policies instead of the failure of cryptographic primitives or protocols. This problem becomes increasingly severe as software systems become more and more complex, and are deployed to manage a large amount of sensitive information and resources organized into sophisticated structures. Identifying discrepancies between policy specifications and their properties (intended function) is crucial because correct implementation and enforcement of policies by applications is based on the premise that the policy specifications are correct. As a result, policy specifications must undergo rigorous verification and validation through systematic testing to ensure that the policy specifications truly encapsulate the desires of the policy authors.

To formally and precisely capture the security properties that access control should adhere to, access control models are usually written to bridge the rather wide gap in abstraction between policy and mechanism. Thus, an access control model provides unambiguous and precise expression as well as reference for design and implementation of security requirements. Techniques are required for verifying whether an access control model is correctly expressed in the access controls policies and whether the properties are satisfied in the model. In practice, the same access control policies may express multiple access control models or express a single model in addition to extra access control constraints outside of the model. Ensuring the conformance of access control models and policies is a nontrivial and critical task.

Started in 2009, CSD developed a prototype system, Access Control Policy Tool (ACPT), which allows a user to compose, verify, test, and generate access control policies.

In FY2012, ACPT was downloaded by 130 users and organizations. CSD performed Beta testing, enhanced the capability of ACPT by adding new policy combine algorithms, applied more stringent and practical user cases to test ACPT's performance, and researched an additional modeling method that is more flexible than the current one used. CSD also produced a new user manual that contains examples and detailed information of ACPT. In addition, CSD published a research paper related to ACPT.

In FY2013, CSD will continue testing, enhance the capability of ACPT by adding environmental variable function, provide model profiles, and resolve compatibility issues between systems used by ACPT. CSD will also update ACPT from users' feedbacks and suggestions.

This project is expected to:

* Provide generic paradigm and framework of access control model/property conformance testing;

* Provide templates for specifying access control rules in popular access control models such as Attribute Based, Multilevel,

and Workflow models;

* Provide tools or services for checking the security and safety of access control implementation, policy combination, and XACML policy generation;

* Promote (or accelerate) the adoption of combinatorial testing for large-system (such as access control system) testing; and

* Assist system architects, security administrators, and security managers whose expertise is related to access control in managing their systems, and to learn the limitations and practical approaches for their applications.

http://csrc.nist.gov/groups/SNS/acpt/

Contacts:
Dr. Vincent Hu
(301) 975-4975
vhu@nist.gov

Mr. Rick Kuhn
(301) 975-3337
kuhn@nist.gov

→ Metrics for Evaluation of Access Control Systems

Access control (AC) systems come with a wide variety of features and administrative capabilities, and the operational impact can be significant. In particular, this impact can pertain to administrative and user productivity, as well as to the organization's ability to perform its mission. Therefore, it is reasonable to use a quality metric to verify the mechanical properties of AC systems. Features that influence the development of this metric are: 1) administration is the main consideration of cost; 2) enforcement capabilities are the requirements for AC applications; 3) the performance is the major factor for the AC usability; and 4) support functions allow an AC system to utilize and connect to related technologies so as to enable more efficient integration with network and host service functions. This project provides a metric for the evaluation of AC systems based on the features of administration, enforcement, performance, and support of AC properties.

The ability of an organization to enforce its access policies determines the degree to which its data may be protected and shared among its user community. The focus on sharing and protecting information is becoming increasingly acute for many organizations. Unfortunately, when it comes to AC systems, one size does not fit all. The quality of administrative capabilities has an impact on administrative cost, user downtime between administrative events, and the abilities of users to perform their duties, as well as the overall security posture of the enterprise. Currently no well-accepted metrics exist for measuring the effectiveness or functional quality of an AC system.

The purpose of this project is to provide federal agencies with background information on access control properties, and to help agencies improve the evaluation of their AC systems. This project provides information of the administration, enforcement, performance, and support properties of AC mechanisms that are embedded in each AC system. Properties discussed in this project extend to the information in NISTIR 7316, *Assessment of Access Control Systems*, which demonstrates the fundamental concept of policy, models, and mechanisms of AC systems.

In FY2012, CSD completed the writing of NISTIR 7874, *Guidelines for Access Control System Evaluation Metrics*. NISTIR 7874 includes detailed items for AC system properties, as well as examples to demonstrate how to use the metric in evaluating and comparing capabilities for AC systems, which can be applied to application or research environments.

CSD expects that this project will:

* Provide detailed information on the evaluation of AC systems, including policies, models, and mechanism for AC system researchers;
* Help security policy makers and system administrators in planning and improving their current and extended future AC systems;
* Provide information for AC system developers in the consideration of architecture, requirements, and performance of an AC system; and
* Provide reference information for AC system-related standards.

Contacts:

Dr. Vincent Hu
(301) 975-4975
vhu@nist.gov

Mr. David Ferraiolo
(391) 975-3046
david.ferraiolo@nist.gov

Mr. Rick Kuhn
(301) 975-3337
kuhn@nist.gov

Advanced Security Testing and Measurements

→ Security Automation and Vulnerability Management

Security automation harmonizes the vast amount of IT product data into coherent, comparable information streams to achieve situational awareness that informs timely and active management of diverse IT systems. Through the creation of flexible, open standards and international recognition, security automation will result in IT infrastructure interoperability, broad acceptance, and adoption, and will create opportunities for innovation.

Security Content Automation Protocol (SCAP) / SCAP Specification Development

To support the overarching security automation vision, it is necessary to have both trusted information and a standardized means to store and share it. Through close work with its government and industry partners, NIST has developed the Security Content Automation Protocol (SCAP) to provide the standardized technical mechanisms to share information between systems. Through the National Vulnerability Database (NVD) and the National Checklist Program (NCP), NIST is providing relevant and important information in the areas of vulnerability and configuration management. Combined, SCAP and the programs that leverage it are moving the information assurance industry towards being able to standardize communications, collect and store relevant data in standardized formats, and provide automated means for the assessment and remediation of systems for both vulnerabilities and configuration compliance.

SCAP is a suite of specifications that use Extensible Markup Language (XML) to standardize the format and nomenclature by which security software products communicate information about software flaws and security configurations. SCAP includes software flaw and security configuration standard reference data, also known as SCAP content. This reference data is provided by the NVD (http://nvd.nist.gov/).

SCAP is a multipurpose protocol that supports automated vulnerability checking, technical control compliance activities, and security measurement. The U.S. government, in cooperation with academia and private industry, is adopting SCAP and encourages its use in support of security automation activities and initiatives.

At the end of September 2012, SP 800-126 Revision 2, *The Technical Specification for the Security Content Automation Protocol (SCAP): SCAP Version 1.2,* was issued as final (http://csrc.nist.gov/publications/ nistpubs/800-126-rev2/SP800-126r2.pdf). This document describes the 11 component specifications composing SCAP:

* Languages:
  ○ Extensible Configuration Checklist Description Format (XCCDF), a language for authoring security checklists/benchmarks and for reporting results of evaluating them;
  ○ Open Vulnerability and Assessment Language (OVAL), a language for representing system configuration information, assessing machine state, and reporting assessment results; and
  ○ Open Checklist Interactive Language (OCIL), a language for representing checks that collect information from people or from existing data stores made by other data collection efforts;

* Reporting Formats:
  ○ Asset Reporting Format (ARF), a format for expressing the transport format of information about assets and the relationships between assets and reports; and
  ○ Asset Identification (AI), a format for uniquely identifying assets based on known identifiers and/or known information about the assets;

* Enumerations:
  ○ Common Platform Enumeration (CPE), a nomenclature and dictionary of hardware, operating systems, and applications;
  ○ Common Configuration Enumeration (CCE), a nomenclature and dictionary of software security configurations; and
  ○ Common Vulnerabilities and Exposures (CVE), a nomenclature and dictionary of security-related software flaws;

* Measurement and Scoring Systems:
  ○ Common Vulnerability Scoring System (CVSS), a specification for measuring the relative severity of software flaw vulnerabilities; and
  ○ Common Configuration Scoring System (CCSS), a specification for measuring the relative severity of system security configuration issues; and

* Integrity:
  ○ Trust Model for Security Automation Data (TMSAD), a specification for using digital signatures in a common trust model applied to security automation specifications.

SCAP is being widely adopted by major software and hardware manufacturers and has become a significant component of information security management and governance programs. The protocol is expected to evolve and expand in support of the growing need to define and measure effective security controls, assess and monitor ongoing aspects of information security, remediate noncompliance, and successfully manage systems in accordance with the Risk Management Framework described in SP 800-53.

Currently, CSD is leveraging SCAP in multiple areas, both to support its own mission and to enable other agencies and private sector entities to meet their goals. For CSD, SCAP is a critical component of the SCAP Validation Program, the National Vulnerability Database (NVD), and the National Checklist Program.

Contact:
Mr. David Waltermire
(301) 975-3390
david.waltermire@nist.gov

National Vulnerability Database (NVD)

The National Vulnerability Database (NVD) is the U.S. government repository of standards-based vulnerability management reference data. The NVD provides information regarding security vulnerabilities and configuration settings, vulnerability impact metrics, technical assessment methods, and references to remediation assistance and IT product identification data. The NVD reference data supports security automation efforts based on the Security Content Automation Protocol (SCAP). As of September 2012, the NVD contained the following resources:

* Over 53,000 vulnerability advisories with an average of 8 new vulnerabilities added daily;
* 46 SCAP-expressed checklists containing thousands of low-level security configuration checks that can be used by SCAP-validated security products to perform automated evaluations of system state;
* 156 non-SCAP security checklists (e.g., English prose guidance and configuration scripts);
* 222 U.S. Computer Emergency Readiness Team (USCERT) alerts, 2,645 US-CERT vulnerability summaries, and 8,140 SCAP machine-readable software flaw checks;
* Product dictionary with 62,729 operating system, application, and hardware name entries; and
* 38,083 vulnerability advisories translated into Spanish.

NVD is hosted and maintained by NIST and is sponsored by the Department of Homeland Security's National Cyber Security Division.

NVD's effective reach has been extended by the use of SCAP data by commercial security products deployed in thousands of organizations worldwide. Increased adoption of SCAP is evidenced by the increasing demand for NVD Extensible Markup Language (XML) data feeds and SCAP-expressed content from the NVD website. Concerted outreach efforts over the last year have resulted in an increase in the number of vendors providing SCAP-expressed content.

NVD continues to play a pivotal role in the Payment Card Industry (PCI) efforts to mitigate vulnerabilities in credit card systems. PCI mandates the use of NVD vulnerability severity scores in measuring the risk to payment card servers worldwide and for prioritizing vulnerability patching. PCI's use of NVD severity scores helps enhance credit card transaction security and protects consumers' personal information.

Throughout FY2012, NVD continued to provide access to vulnerability reference data and security checklists. NVD deployed an enhanced checklist submission web interface and a web service checklist submission capability. Additionally, the NVD now hosts a SCAP Content Validation Tool that can be used by creators of SCAP content to ensure that their SCAP content packages conform to Special Publication (SP) 800-126, *The Technical Specification for the Security Content Automation Protocol (SCAP): SCAP Version 1.2*, guidelines. Finally, NVD now supports automated SCAP content generation from the Common Vulnerabilities and Exposures (CVE) vulnerability data feed. NVD data is a fundamental component of CSD's security automation infrastructure and is substantially increasing the security of networks worldwide. CSD plans to expand and improve the NVD in FY2013.

http://nvd.nist.gov

Contact:
Mr. Harold Booth
(301) 975-8441
harold.booth@nist.gov

Incident Handling Automation

In recent years, security threats to digital systems have become more prevalent and more sophisticated. While some security threats are generic in nature, others are targeted at specific organizations, assets, and missions. Although computer security defenses may forestall many threats, not all can be prevented, and organizations must therefore develop incident handling capabilities. Incident handling encompasses a variety of tasks ranging from preparation prior to an incident to timely detection and analysis of an incident to recovery and repair from the effects of an incident to post-incident learning and improvement. These tasks need to be performed both internally within specific organizations and externally via coordination across teams of collaborating organizations.

In the past year, NIST worked with the Department of Homeland Security's United States Computer Emergency Readiness Team (US-CERT) to develop Revision 2 of NIST Special Publication (SP) 800-61, *Computer Security Incident Handling Guide*. This document provides guidance on developing incident handling capabilities. The document explains the nature of incidents, explains the incident handling process, explains the structure and operation of Computer Security Incident Response Teams (CSIRTs), and provides guidance on handling an incident and coordinating with other organizations.

SP 800-61 focuses primarily on manual (human) processes for incident handling and the effective use of human judgment, guided by applicable regulation and law, regarding which incident-related information is significant and which incident-related information may be shared. The growing volume of security threats, however, is driving the need for a more agile incident-handling framework that can operate at differing scales and speeds as required.

Working in concert with the Department of Homeland Security, NIST is expanding existing incident handling guidance to enable coordinated information sharing across disparate CSIRTs operating at differing scales and speeds. This work will include the analysis of standardized incident handling data models and the incorporation of these data models, as appropriate, into both CSIRT information sharing processes as well as incident/threat knowledge repositories. This work will describe how mature CSIRTs may operate in a diverse information-sharing network with both operational and strategic CSIRTs, as well as industry knowledge repositories. This may include selective use of security automation where applicable.

In FY2013, this work will develop draft SP 800-150, *Coordinated Computer Security Incident Handling Guidance*.

Contacts:
Mr. Lee Badger
(301) 975-3176
lee.badger@nist.gov

Mr. David Waltermire
(301) 975-3390
david.waltermire@nist.gov

United States Government Configuration Baseline (USGCB) / FDCC Baselines

The United States Government Configuration Baseline (USGCB) initiative creates security configuration baselines for information technology (IT) products widely deployed across the federal agencies. The project evolved from the Federal Desktop Core Configuration (FDCC) mandate originally described in a March 2007 memorandum from the U.S. White House Office of Management and Budget (Memorandum M-07-11). USGCB helps to improve information security and reduce overall IT operating costs by providing commonly accepted security configurations for major operating systems.

Through the National Checklist Program described in Special Publication (SP) 800-70, *National Checklist Program for IT Products: Guidelines for Checklist Users and Developers*, a baseline submitter may express interest in submitting a candidate for use in the USGCB program. CSD works with the Federal CIO Council's Technology Infrastructure Subcommittee (TIS) to consider the candidate, recommend any changes to the baseline, and coordinate implementation at federal agencies.

On behalf of the TIS, NIST reviews the SCAP-expressed checklist to ensure that it complies with the appropriate specifications and to ensure that the benchmark properly assesses the intended security configuration. Where possible, virtual images of target architectures are provided to assist agencies with testing the content for suitability in their environment. For example, Windows desktop virtual hard drive images (VHDs) and Group Policy Object (GPO) files or Red Hat Enterprise Linux Kickstart Configuration Scripts may be provided, enabling users to replicate the target test environment for local validation. Any identified issues may be posted to the USGCB and FDCC mail aliases.

CSD provides ongoing support for the USGCB automation content, including the creation of patch updates, assisting USGCB users in continuously monitoring and assessing security compliance of information systems. This ongoing monitoring element supports the Risk Management Framework described in SP 800-37 Revision 1, *Guide for Applying the Risk Management Framework to Federal Information Systems: A Security Life Cycle Approach.*

The USGCB Program will continue in FY2013 to provide ongoing maintenance of the baseline artifacts and to consider additional applicable platforms.

Contact:
Mr. Stephen Quinn
(301) 975-6967
stephen.quinn@nist.gov

National Checklist Program (NCP)

There are many threats to information technology (IT), ranging from remotely launched network service exploits to malicious code spread through infected emails, websites, and downloaded files. Vulnerabilities in IT products are discovered daily, and many ready-to-use exploitation techniques are widely available on the Internet. Because IT products are often intended for a wide variety of audiences, restrictive security configuration controls are usually not enabled by default. As a result, many out-of-the box IT products are immediately vulnerable. In addition, identifying a reasonable set of security settings that achieve balanced risk management is a complicated, arduous, and time-consuming task, even for experienced system administrators.

To facilitate development of security configuration checklists for IT products and to make checklists more organized and usable, NIST established the National Checklist Program (NCP) in furtherance of its statutory responsibilities under the Federal Information Security Management Act (FISMA) of 2002, Public Law 107-347, and also under the Cyber Security Research and Development Act, which tasks NIST to "develop, and revise as necessary, a checklist setting forth settings and option selections that minimize the security risks associated with each computer hardware or software system that is, or is likely to become widely used within the federal government." In February 2008, revised Part 39 of the Federal Acquisition Regulation (FAR) was published. Paragraph (d) of section 39.101 states, "In acquiring information technology, agencies shall include the appropriate IT security policies and requirements, including use of common security configurations available from the NIST website at http://checklists.nist.gov. Agency contracting officers should consult with the requiring official to ensure the appropriate standards are incorporated." In Memorandum M08-22, Office of Management and Budget (OMB) mandated the use of SCAP-validated products for continuous monitoring of Federal Desktop Core Configuration (FDCC) compliance. The NCP strives to encourage and make simple agencies' compliance with these mandates.

The goals of the NCP are to:

* Facilitate development and sharing of checklists by providing a formal framework for checklist developers to submit checklists to NIST;
* Provide guidance to developers to help them create standardized, high-quality checklists that conform to common operations environments;
* Help developers and users by providing guidelines for making checklists better documented and more usable;
* Encourage software vendors and other parties to develop checklists;
* Provide a managed process for the review, update, and maintenance of checklists;
* Provide an easy-to-use repository of checklists; and,
* Encourage the use of automation technologies for checklist application such as SCAP.

There are 234 checklists posted on the website; 39 of the checklists are SCAP-expressed (see section on SCAP above) and can be used with SCAP-validated products. It is anticipated that a minimum of several more SCAP-expressed checklists will be added in FY2013 as contributions come from other federal agencies and product vendors. Organizations can use checklists obtained from the NCP website (http://checklists. nist.gov) for automated security configuration patch assessment. NCP currently hosts SCAP checklists for Internet Explorer 7.0, Internet Explorer 8.0, Microsoft Office 2007, Red Hat Enterprise Linux, Windows 7,

Windows Vista, Windows XP, and other products.

To assist users in identifying automated checklist content, NCP groups checklists into tiers, from Tier I to Tier IV. NCP uses the tiers to rank checklists according to their automation capability. Tier III and IV checklists are considered production-ready and have been validated by the SCAP content validation tool as conforming to the requirements outlined in SP 800-126, *The Technical Specification for the Security Content Automation Protocol (SCAP)*. Tier IV checklists are used in the SCAP Validation Program (see following section for details) when validating SCAP products. Tier III checklists are not presently used in the SCAP Validation Program; however, Tier III checklists should be compatible with SCAP-validated products. Tier II checklists document recommended security settings in a machine-readable, nonstandard format, such as a proprietary format or a product-specific configuration script. Tier I checklists are prose-based and contain no machine-readable content. Users can browse the checklists based on the checklist tier, IT product, IT product category, or authority, and also through a keyword search that searches the checklist name and summary for userspecified terms. The search results show the detailed checklist metadata and a link to any SCAP content for the checklist, as well as links to any supporting resources associated with the checklist.

To assist checklist developers, the NCP provides both manual and automated interfaces to facilitate submission and maintenance processes. The manual interface consists of a web application that guides the submitter through the data entry process to ensure that all of the required information is submitted. In addition, a web service is also available for a fully automated submission. In either case, the submission is validated upon review, and a report is returned to the submitting organization, verifying either acceptance or rejection based on the criteria requirements. For instance, Tier III and Tier IV checklists require validation using the SCAP Content Validation Tool (this tool is available for download via http://scap.nist.gov/revision/1.2/#tools).

The NCP is defined in SP 800-70 Revision 2, *National Checklist Program for IT Products—Guidelines for Checklist Users and Developers*, which can be found at http://csrc.nist.gov/publications/.

http://checklists.nist.gov

Contact:
Mr. Stephen Quinn
(301) 975-6967
stephen.quinn@nist.gov

→ Technical Security Metrics

Security Risk Analysis of Enterprise Networks Using Attack Graphs

At present, computer networks constitute the core component of information technology infrastructures in areas such as power grids, financial data systems, and emergency communication systems. Protection of these networks from malicious intrusions is critical to the economy and security of the nation. Vulnerabilities are regularly discovered in software applications which are exploited to stage cyber attacks. Currently, management of security risk of an enterprise network is more an art than a science. System administrators operate by instinct and experience rather than relying on objective metrics to guide and justify decision making. The objective of this research is to develop a standard model for measuring security of computer networks. A standard model will enable us to answer questions such as "Are we more secure now than yesterday?" or "How does the security of one network configuration compare with another one?" Also, having a standard model to measure network security will allow users, vendors, and researchers to evaluate methodologies and products for network security in a coherent and consistent manner.

The CSD has approached the challenge of network security analysis by capturing vulnerability interdependencies and measuring security in the exact way that real attackers penetrate the network. CSD's methodology for security risk analysis is based on the model of attack graphs. CSD analyzes all attack paths through a network, providing a probabilistic metric of the overall system risk. Through this metric, CSD analyzes trade-offs between security costs and security benefits.

In FY2012, CSD integrated techniques into an attack graph-based security tool called MULVAL. CSD released this as an open source system. CSD published a Springer brief book "Quantitative Security Risk Assessment of Enterprise Networks" on this topic.

CSD also developed a good model "K Zero Day Safety" for measuring security risk of a network against unknown vulnerabilities. CSD prepared a paper "A Network Security Metric for Measuring and Improving Computer Network Resistance to Unknown Vulnerabilities."

In FY2013, CSD plans to apply attack graphs to study the effectiveness of moving target defense. CSD also plans to publish the results as a NIST report and as white papers in conferences and journals.

http://csrc.nist.gov/groups/SNS/security-risk-analysis-enterprise-networks/

Contact:
Dr. Anoop Singhal
(301) 975-4432
anoop.singhal@nist.gov

Using Attack Graphs in Forensic Examinations

Attack graphs are used to compute potential attack paths from a system configuration and known vulnerabilities of a system. Attack graphs can be used to determine known vulnerability sequences that were exploited to launch the attack and help forensic examiners in identifying many potential attack paths. After an attack happens, forensic analysis, including linking evidence with attacks, helps further understand and refine the attack scenario that was launched. Given that there are anti-forensic tools that can obfuscate, minimize, or eliminate attack footprints, forensic analysis becomes harder. In this project, CSD applied attack graphs to forensic analysis. CSD did so by including anti-forensic capabilities into attack graphs, so that the missing evidence can be explained by using longer attack paths that erase potential evidence.

In FY2012, CSD published a paper, "Using Attack Graphs in Forensic Examination," in an IEEE Workshop on Digital Forensics (WSDF 2012). In FY2013, CSD plans to enhance this work by mapping Evidence Graphs to Attack Graphs. CSD also plans to enhance an existing attack graph generation tool to generate evidence graphs for forensic examinations. CSD also plans to publish the results as a NIST report and as papers in conferences and journals.

Contact:
Dr. Anoop Singhal
(301) 975-4432
anoop.singhal@nist.gov

→ Automated Combinatorial Testing

Software developers often encounter failures that result from an unexpected interaction between components. NIST research has shown that most failures are triggered by one or two parameters, and progressively fewer by three, four, or more parameters (see graph on page 69), a relationship that is called the interaction rule. These results have important implications for testing. If all faults in a system can be triggered by a combination of n or fewer parameters, then testing all n-way combinations of parameters can provide very strong fault detection efficiency. These methods are being applied to software and hardware testing for reliability, safety, and security. CSD's focus is on empirical results and real-world problems.

Project highlights for FY2012 included demonstration of 95 percent reduction in test volume with equivalent fault detection for interoperability of Internet web browser software; completion of the first textbook on combinatorial testing; cooperative work with the National Aeronautics and Space Administration (NASA) Independent Verification and Validation (IV&V) Facility demonstrating the effectiveness of combinatorial methods for IV&V of space systems; lectures at conferences and research labs; and leading (jointly with IBM personnel) the IEEE First International Conference on Combinatorial Testing, held with the International Conference on Software Testing.

Tech transfer activities included publication of six technical papers; release of enhanced covering array, test prioritization, and fault location tools; plus seminars and lectures at Carnegie Mellon University, NASA, Loyola College, information technology companies, Indian Institute of Technology, and several conferences.

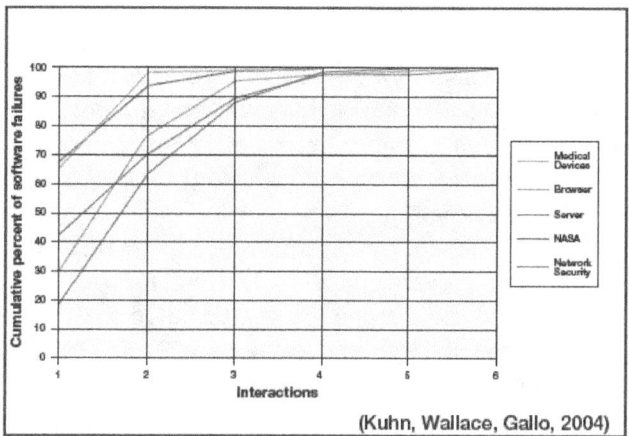

(Kuhn, Wallace, Gallo, 2004)

Plans for FY2013 include a new project with the NASA Independent Verification and Validation (IV&V) Facility to apply combinatorial coverage measurement methods to NASA spacecraft software, following a successful pilot study in 2012; development of new methods and tools for fault location; lectures at conferences and research labs; and significant enhancement of software tools for combinatorial coverage measurement, jointly with Centro Nacional de Metrología, the national metrology institute of Mexico.

http://csrc.nist.gov/groups/SNS/acts/

Contacts:

Mr. Rick Kuhn
(301) 975-3337
kuhn@nist.gov

Dr. Raghu Kacker
(301) 975-2109
raghu.kacker@nist.gov

➔ Hardware Roots of Trust

Modern computing devices consist of various hardware, firmware, and software components at multiple layers of abstraction. Many security and protection mechanisms are currently rooted in software that, along with all underlying components, must be trusted and uncorrupted. A vulnerability in any of those components could compromise the trustworthiness of the security mechanisms that rely upon those components. Stronger security assurances may be possible by grounding security mechanisms in roots of trust.

Roots of trust are highly reliable and secure hardware, firmware, and software components that perform specific, critical security functions. Because roots of trust are inherently trusted, they must be secure by their design. As such, many roots of trust are implemented in hardware so that malware cannot tamper with the functions they provide. Roots of trust provide a firm foundation from which to build security and trust in a system.

NIST's past work on roots of trust has focused on their use to protect fundamental system firmware, commonly known as the Basic Input/Output System (BIOS). NIST has been working with industry on the use of roots of trust to improve the security of BIOS. In FY2011, NIST issued SP 800-147, *BIOS Protection Guidelines*, which provides guidelines on protecting BIOS in laptop and desktop computers. NIST extended these guidelines in FY2012 to cover server-class systems, and released draft NIST SP 800-147B, *BIOS Protection Guidelines for Servers*, for public comment. NIST is also developing guidelines on the use of roots of trust to detect unauthorized changes to BIOS and BIOS configuration settings. In FY2012, NIST released draft SP 800-155, *BIOS Integrity Measurement Guidelines*, which provides guidelines on mechanisms that measure and report the state of BIOS.

NIST will continue its efforts to secure BIOS and other critical firmware in FY2013. CSD will finalize the BIOS protection and measurement guidelines (i.e., SP 800-147B and SP 800-155), and will develop new guidelines for protection of firmware in computer add-on cards. Future efforts will explore methods to extend trust in the security of BIOS to provide greater assurance of the security of the operating system and applications.

A new focus area for NIST's roots of trust research will be in mobile devices. For the past year, NIST has been working with government and industry partners on guidelines for hardware-rooted security features in mobile devices. These guidelines will focus on device integrity, isolation, and protected storage features that are supported by roots of trust. A draft of these guidelines will be released for public comment in early FY2013.

Contact:
Mr. Andrew Regenscheid
(301) 975-5155
andrew.regenscheid@nist.gov

→ Continuous Monitoring Reference Model

In September 2010, the Department of Homeland Security (DHS) released the *Continuous Asset Evaluation, Situational Awareness and Risk Scoring* (CAESARS) Reference Architecture Report. This report identifies commonality and strengths in the custom approaches used by civilian agencies to provide solutions that enable the continuous monitoring of IT systems. This report identifies "essential functional components of a security risk scoring system, independent of specific technologies, products, or vendors." It describes the use of security automation specifications, such as the Security Content Automation Protocol (SCAP), to enable continuous monitoring solutions.

In October 2010, the Federal Chief Information Officer Council's Information Security and Identity Management Committee's (ISIMC) subcommittee on Continuous Monitoring and Risk Scoring saw the need to create a technical initiative to expand upon the CAESARS architecture to better scale it to large enterprises (e.g., the entire U.S. government). A team of researchers from the National Security Agency's (NSA) Information Assurance Directorate (IAD), the DHS Federal Network Security CAESARS team, and NIST's Information Technology Laboratory (ITL) worked together to respond to this need. The draft CAESARS Framework Extension (FE) described by the NIST Interagency Report (NISTIR) 7756, *CAESARS Framework Extension: An Enterprise Continuous Monitoring Technical Reference Architecture*, is the output of this collaboration.

This report presents an enterprise continuous monitoring technical reference architecture that extends the framework provided by the DHS's CAESARS architecture. The primary goal of this effort is to facilitate enterprise continuous monitoring by presenting a reference architecture that enables organizations to aggregate collected data from across a diverse set of security tools, analyze that data, perform scoring, enable user queries, and provide overall situational awareness. In support of this goal, the CAESARS-FE reference architecture describes additional functionality, provides more granularity within subsystem specifications, and further leverages security automation efforts. The model design is focused on enabling organizations to realize this capability by leveraging their existing security tools based on the use of open, consensus-based standards, to avoid complicated and resource-intensive custom tool integration efforts.

The data exchange and functional requirements in CAESARS-FE and referenced specifications provide organizations with much of the information needed to bring together diverse security products, and use those products to compose a hierarchical data aggregation model that supports a large variety of continuous monitoring consumers from both the security disciplines and general information technology (IT) management domains. CAESARS-FE minimally defines the required functionality so that security tool vendors can cost-effectively participate, while ensuring a necessary level of interoperability between vendor products.

To advance the state of the art in continuous monitoring capabilities and to further interoperability within commercially available tools, CSD is working within the international standards development community to establish working groups and to author and comment on emerging technical standards in this area. The CAESARS-FE reference architecture is expected to evolve as greater consensus is developed around interoperable, standards-based approaches that enable continuous monitoring of IT systems.

Contact:
Mr. David Waltermire
(301) 975-3390
david.waltermire@nist.gov

➔ Universal Credential Revocation

Through NIST's identity-related project and hosting the IDTrust Symposium, credential revocation has emerged as one of the key gaps in progressing secure authentication online. In federated environments, credential revocation has traditionally been managed by the credential issuer. In an effort to improve credential revocation mechanisms across federations and effectively mitigate credential misuse, NIST researchers are exploring the broader scope of credential revocation, where all parties contribute to and participate in credential revocation. In this model, service providers give feedback on a credential reliability score based on detected credential misuse. The credential holder and Identity Provider, on the other hand, receive feedback notifications and are able to immediately suspend or revoke the credential should the score reach an unacceptable level. Lastly, other federation services can consult scores and status to determine the suitability of a presented credential with an associated reliability score.

In FY2012, NIST scientists published an initial draft NISTIR 7817 titled *A Credential Reliability and Revocation Model* that describes and classifies the different types of identity providers serving federations. For each classification, the document identifies perceived improvements or gaps when the credentials are used in authentication services and recommends countermeasures to eliminate some of the identified gaps. With the countermeasures as the basis, the document suggests a Universal Credential Reliability and Revocation Services (URRS) model that strives to improve authentication services for federations where all parties contribute to and participate in credential revocation and where the service can be tailored to the type of (classification of) identity providers it accommodates.

After the public comment period, NISTIR 7817 was further revised to accommodate the comments and final approval is anticipated in FY2013.

Contact:
Ms. Hildegard Ferraiolo
(301) 975-6972
hildegard.ferraiolo@nist.gov

Donna Dodson
2012 Annual List of Top Cybersecurity Leaders: 10 Most Influential People in Government Infosec

Award Background: Identifying the top 10 influential people in government information security for the coming year, *in many respects, is an easy undertaking: there are scores of individuals to choose from among potential candidates. Small wonder, a number of industrious practitioners and leaders work diligently to help safeguard government IT.*

Creating IT security guidance requires strong collaboration between NIST and its stakeholders, and Donna Dodson is the facilitator of much of that interaction. Donna Dodson knows how to engage successfully with the White House; civilian, defense and intelligence agencies; and industry on cybersecurity, keeping these constituencies focused on the issues. She is heavily invested in initiatives to secure cloud and mobile computing, the Smart Grid, and supply chain.

Kevin Stine
2012 WEDI Award of Merit

Kevin Stine was designated by the Workgroup for Electronic Data Interchange (WEDI) Board of Directors as a recipient of the 2012 WEDI Award of Merit. WEDI, established in 1991 in response to a challenge from then Secretary of Health and Human Services, Louis Sullivan, MD, brings together a consortium of leaders within the healthcare industry to identify practical strategies for reducing administrative costs in healthcare through the implementation of electronic data interchange. This award recognizes individuals who have contributed in a meaningful way to the success of WEDI and/or WEDI Strategic National Implementation Process (SNIP) programs and activities through their volunteer commitment and talents.

Dr. Ron Ross
Distinguished Practitioner Award 28th Annual Computer Security Applications Conference

Dr. Ross received the Distinguished Practitioner Award at the 28th Annual Computer Security Applications Conference. This award is presented to an individual who has demonstrated a continuing, vital, and influential contribution to the field of information security.

SC Magazine's Influential IT Security Minds in 2012
Dr. Ross was also recognized by SC Magazine as one of the Influential IT Security Minds in 2012. This recognizes six luminaries who represent the highest degree of professionalism in the security space, industry veterans who stand out for their skills, managerial prowess, insight and advocacy.

2012 Commencement Address
George Washington University
School of Engineering and Applied Science

Dr. Ross also delivered the 2012 Commencement Address at the George Washington University's School of Engineering and Applied Science.

This list of publications (below) were released by the Computer Security Division in FY2012 (from October 1, 2011 to September 30, 2012). In addition, the only draft publication that was approved final during FY2012 was SP 800-61 Rev. 2.

Type & Number	Publication Title	Draft Released Date
DRAFT PUBLICATIONS		
Federal Information Processing Standards (FIPS)		
FIPS 201-2	Personal Identity Verification (PIV) of Federal Employees and Contractors	July 2012
FIPS 186-3	Proposed Change Notice for Digital Signature Standard (DSS)	April 2012
Special Publications (SPs)		
SP 800-155	BIOS Integrity Measurement Guidelines	Dec. 2011
SP 800-152	A Profile for U.S. Federal Cryptographic Key Management Systems (CKMS)	Aug. 2012
SP 800-147B	BIOS Protection Guidelines for Servers	July 2012
SP 800-130	A Framework for Designing Cryptographic Key Management Systems	Apr. 2012
SP 800-124 Rev. 1	Guidelines for Managing and Securing Mobile Devices in the Enterprise	July 2012
SP 800-117 Rev. 1	Guide to Adopting and Using the Security Content Automation Protocol (SCAP) Version 1.2	Jan. 2012
SP 800-94 Rev. 1	Guide to Intrusion Detection and Prevention Systems (IDPS)	July 2012
SP 800-90C	Recommendation for Random Bit Generator (RBG) Constructions	Sept. 2012
SP 800-90B	Recommendation for the Entropy Sources Used for Random Bit Generation	Sept. 2012
SP 800-88 Rev. 1	Guidelines for Media Sanitization	Sept. 2012
SP 800-83 Rev. 1	Guide to Malware Incident Prevention and Handling for Desktops and Laptops	July 2012
SP 800-76-2	Biometric Data Specification for Personal Identity Verification	July 2012
SP 800-61 Rev. 2 *(This draft document was approved as final in August 2012.)*	Computer Security Incident Handling Guide	Feb. 2012
SP 800-56A Rev. 2	Recommendation for Pair-Wise Key-Establishment Schemes Using Discrete Logarithm Cryptography	Aug. 2012
SP 800-53 Rev. 4	Security and Privacy Controls for Federal Information Systems and Organizations	Feb. 2012
SP 800-40 Rev. 3	Guide to Enterprise Patch Management Technologies	Sept. 2012
NIST Interagency Reports (NISTIRs)		
NISTIR 7848	Specification for the Asset Summary Reporting Format 1.0	May 2012
NISTIR 7831	Common Remediation Enumeration (CRE) Version 1.0	Dec. 2011
NISTIR 7823	Advanced Metering Infrastructure Smart Meter Upgradeability Test Framework	July 2012
NISTIR 7817	A Credential Reliability and Revocation Model for Federated Identities	Jan. 2012
NISTIR 7800	Applying the Continuous Monitoring Technical Reference Model to the Asset, Configuration, and Vulnerability Management Domains	Jan. 2012
NISTIR 7799	Continuous Monitoring Reference Model Workflow, Subsystem, and Interface Specifications	Jan. 2012
NISTIR 7756	CAESARS Framework Extension: An Enterprise Continuous Monitoring Technical Reference Architecture	Jan. 2012
NISTIR 7622	Notional Supply Chain Risk Management Practices for Federal Information Systems	March 2012
NISTIR 7511 Rev. 3.04	Security Content Automation Protocol (SCAP) Version 1.0 Validation Program Test Requirements	Sept. 2012

PUBLICATIONS APPROVED AS FINAL

Federal Information Processing Standards (FIPS)

FIPS Number	Publication Title	Approval Date
FIPS 180-4	Secure Hash Standard (SHS)	March 2012

Special Publications (SPs)

SP Number	Publication Title	Approval Date
SP 500-295	Conformance Testing Methodology for ANSI/NIST-ITL 1-2011, Data Format for the Interchange of Fingerprint, Facial & Other Biometric Information (Release 1.0)	Aug. 2012
SP 800-153	Guidelines for Securing Wireless Local Area Networks (WLANs)	Feb. 2012
SP 800-146	Cloud Computing Synopsis and Recommendations	May 2012
SP 800-144	Guidelines on Security and Privacy in Public Cloud Computing	Dec. 2011
SP 800-135 Rev. 1	Recommendation for Existing Application-Specific Key Derivation Functions	Dec. 2011
SP 800-121 Rev. 1	Guide to Bluetooth Security	June 2012
SP 800-107 Rev. 1	Recommendation for Applications Using Approved Hash Algorithms	Aug. 2012
SP 800-90 A	Recommendation for Random Number Generation Using Deterministic Random Bit Generators	Jan. 2012
SP 800-67 Rev. 1	Recommendation for the Triple Data Encryption Algorithm (TDEA) Block Cipher	Jan. 2012
SP 800-63-1	Electronic Authentication Guideline	Dec. 2011
SP 800-61 Rev. 2	Computer Security Incident Handling Guide	Aug. 2012
SP 800-57 Part 1	Recommendation for Key Management: Part 1: General (Revision 3)	July 2012
SP 800-56 C	Recommendation for Key Derivation through Extraction-then-Expansion	Nov. 2011
SP 800-30 Rev. 1	Guide for Conducting Risk Assessments	Sept. 2012

NIST Interagency Reports (NISTIRs)

NISTIR Number	Publication Title	Approval Date
NISTIR 7877	BioCTS 2012: Advanced Conformance Test Architectures and Test Suites for Biometric Data Interchange Formats and Biometric Information Records	Sept. 2012
NISTIR 7874	Guidelines for Access Control System Evaluation Metrics	Sept. 2012
NISTIR 7870	NIST Test Personal Identity Verification (PIV) Cards	July 2012
NISTIR 7864	The Common Misuse Scoring System (CMSS): Metrics for Software Feature Misuse Vulnerabilities	July 2012
NISTIR 7816	2011 Computer Security Division Annual Report	Mar. 2012

ITL Security Bulletins (produced by Computer Security Division)

Release Date	Title
September 2012	Revised Guide Helps Organizations Handle Security Related Incidents
August 2012	Security of Bluetooth Systems and Devices: Updated Guide Issued By the National Institute of Standards and Technology (NIST)
July 2012	Preparing for and Responding to Certification Authority Compromise and Fraudulent Certificate Issuance
June 2012	Cloud Computing: A Review of Features, Benefits, and Risks, and Recommendations for Secure, Efficient Implementations
May 2012	Secure Hash Standard: Updated Specifications Approved and Issued as Federal Information Processing Standard (FIPS) 180-4

Release Date	Title
March 2012	Guidelines for Improving Security and Privacy in Public Cloud Computing
February 2012	Guidelines for Securing Wireless Local Area Networks (WLANS)
January 2012	Advancing Security Automation and Standardization: Revised Technical Specifications Issued for the Security Content Automation Protocol (SCAP)
December 2011	Revised Guideline for Electronic Authentication of Users Helps Organizations Protect the Security of Their Information Systems
October 2011	Continuous Monitoring of Information Security: An Essential Component of Risk Management

These ITL Security Bulletins can be accessed by visiting the Computer Security Resource Center (CSRC) ITL Security Bulletins page. The URL for this page is: http://csrc.nist.gov/publications/PubsITLSB.html.

Abstracts for Publications Released in FY2012

→ Federal Information Processing Standards (FIPS)

Draft FIPS 201-2, *Personal Identity Verification (PIV) of Federal Employees and Contractors*

This standard specifies the architecture and technical requirements for a common identification standard for federal employees and contractors. The overall goal is to achieve appropriate security assurance for multiple applications by efficiently verifying the claimed identity of individuals seeking physical access to federally controlled government facilities and electronic access to government information systems. The standard contains the minimum requirements for a federal personal identity verification system that meets the control and security objectives of Homeland Security Presidential Directive-12 (HSPD-12), including identity proofing, registration, and issuance. The standard also provides detailed specifications that will support technical interoperability among PIV systems of federal departments and agencies. It describes the card elements, system interfaces, and security controls required to securely store, process, and retrieve identity credentials from the card. The physical card characteristics, storage media, and data elements that make up identity credentials are specified in this standard.

Other requirements of draft FIPS 201-2 are specified in the following Special Publications (SPs):

* SP 800-73, *Interfaces for Personal Identity Verification* (interfaces and card architecture for storing and retrieving identity credentials from a smart card);

* SP 800-76, *Biometric Data Specification for Personal Identity Verification* (interfaces and data formats of biometric information);

* SP 800-78, *Cryptographic Algorithms and Key Sizes for Personal Identity Verification* (requirements for cryptographic algorithms);

* SP 800-79, *Guidelines for the Accreditation of Personal Identity Verification Card Issuers* (requirements for the accreditation of the PIV Card issuers);

* SP 800-87, *Codes for the Identification of Federal and Federally-Assisted Organizations* (unique organizational codes for federal agencies);

* SP 800-96, *PIV Card to Reader Interoperability Guidelines* (requirements for card readers);

* SP 800-156, *Representation of PIV Chain-of-Trust for Import and Export* (format for encoding the chain-of-trust for import and export); and

* SP 800-157, *Guidelines for Personal Identity Verification (PIV) Derived Credentials* (requirements for issuing PIV derived credentials).

Draft FIPS 201-2 does not specify access control policies or requirements for federal departments and agencies.

Contacts:

Ms. Hildegard Ferraiolo   Mr. David Cooper
hildegard.ferraiolo@nist.gov david.cooper@nist.gov

Draft FIPS 186-3, *Proposed Change Notice for Digital Signature Standard (DSS)*

FIPS 186-3, *Digital Signature Standard (DSS)*, specifies three techniques for the generation and verification of digital signatures that can be used for the protection of data: the Digital Signature Algorithm (DSA), the Elliptic Curve Digital Signature Algorithm (ECDSA), and the Rivest-Shamir-Adelman (RSA) algorithm. FIPS 186-3 is used in conjunction with the hash functions specified in FIPS 180-4, *Secure Hash Standard (SHS)*.

The following revisions to FIPS 186-3 are proposed:

(1) The Use of Random Bit/Number Generators;

(2) Definition Clarification; and

(3) The Reuse of a Prime Number Generation Seed for RSA Key Pair Generation.

This standard specifies a suite of algorithms that can be used to generate a digital signature. Digital signatures are used to detect unauthorized modifications to data and to authenticate the identity of the signatory. In addition, the recipient of signed data can use a digital signature as evidence in demonstrating to a third party that the signature was, in fact, generated by the claimed signatory. This is known as non-repudiation, since the signatory cannot easily repudiate the signature at a later time.

Contact:
Ms. Elaine Barker
elaine.barker@nist.gov

FIPS 180-4, *Secure Hash Standard (SHS)*

This standard specifies hash algorithms that can be used to generate digests of messages. The digests are used to detect whether messages have been changed since the digests were generated.

Contacts:
Ms. Shu-jen Chang   Ms. Elaine Barker
shu-jen.chang@nist.gov   elaine.barker@nist.gov

→ Special Publications

SP 500-295, *Conformance Testing Methodology for ANSI/NIST-ITL 1-2011, Data Format for the Interchange of Fingerprint, Facial & Other Biometric Information (Release 1.0)*

Conformance testing measures whether an implementation faithfully implements the technical requirements defined in a standard. Conformance testing provides developers, users, and purchasers with increased levels of confidence in product quality and increases the probability of successful interoperability. NIST's Information Technology Laboratory sponsored the development of a conformance testing methodology for ANSI/NIST-ITL 2011, *Data Format for the Interchange of Fingerprint, Facial & Other Biometric Information* (AN-2011) under the NIST/ITL Conformance Testing Methodology Working Group. This testing methodology supports the development of conformance test tools designed to test implementations of AN-2011 transactions and promotes biometrics conformity assessment efforts. The first release includes comprehensive tables of AN-2011 requirements and test assertions for a set of supported AN-2011 Record Types. The tables of requirements and assertions indicate which assertions apply to the traditional encoding format, the National Information Exchange Model (NIEM)-compliant encoding format, or both encoding formats. The testing methodology makes use of specific test assertion syntax to clearly define the assertions associated with each requirement.

Contacts:
Mr. Fernando Podio   Mr. Dylan Yaga
fernando@nist.gov   dylan.yaga@nist.gov

Mr. Christofer McGinnis
christofer.mcginnis@nist.gov

Draft SP 800-155, *BIOS Integrity Measurement Guidelines*

This document outlines the security components and security guidelines needed to establish a secure Basic Input/Output System (BIOS) integrity measurement and reporting chain. Unauthorized modification of BIOS firmware constitutes a significant threat because of the BIOS's unique and privileged position within the PC architecture. The document focuses on two scenarios: detecting changes to the system BIOS code stored on the system flash, and detecting changes to the system BIOS configuration. The document is intended for hardware and software vendors that develop products that can support secure BIOS integrity measurement mechanisms, and may also be of use for organizations developing enterprise procurement or deployment strategies for these technologies.

Contact:
Mr. Andrew Regenscheid
andrew.regenscheid@nist.gov

SP 800-153, *Guidelines for Securing Wireless Local Area Networks (WLANs)*

A wireless local area network (WLAN) is a group of wireless networking devices within a limited geographic area, such as an office building, that exchange data through radio communications. The security of each WLAN is heavily dependent on how well each WLAN component—including client devices, access points (APs), and wireless switches—is secured throughout the WLAN life cycle, from initial WLAN design and deployment through ongoing maintenance and monitoring. The purpose of this publication is to help organizations improve their WLAN security by providing recommendations for WLAN security configuration and monitoring. This publication supplements other NIST publications by consolidating and strengthening their key recommendations.

Contact:
Mr. Murugiah Souppaya
murugiah.souppaya@nist.gov

Draft SP 800-152, *A Profile for U.S. Federal Cryptographic Key Management Systems (CKMS)*

This publication is being developed for use by federal agencies and contractors when designing, implementing, procuring, installing, configuring, and operating a CKMS. This Profile will be based on (draft) Special Publication 800-130, *A Framework for Designing Cryptographic Key Management Systems*. The framework covers topics that should be considered by a product or system designer when designing a CKMS and specifies requirements for the design and its documentation. The Profile, however, will cover not only a CKMS design, but also its procurement, installation, management, and operation throughout its lifetime. Requirements will, therefore, be placed not only on a CKMS product or system, but also on people (procurement officials, installers, managers, and operators) while performing specific tasks involving the CKMS.

Contact:
Ms. Elaine Barker
elaine.barker@nist.gov

Draft SP 800-147B, *BIOS Protection Guidelines for Servers*

Modern computers rely on fundamental system firmware, commonly known as the system Basic Input/Output System (BIOS), to facilitate the hardware initialization process and transition control to the operating system. Unauthorized modification of BIOS firmware by malicious software constitutes a significant threat because of the BIOS's unique and privileged position within the PC architecture. The guidelines in this document include requirements on servers to mitigate the execution of malicious or corrupt BIOS code. They apply to BIOS firmware stored in the BIOS flash, including the BIOS code, the cryptographic keys that are part of the root of trust for update, and static BIOS data. This guide is intended to provide server platform vendors with recommendations and guidelines for a secure BIOS update process.

Contact:
Mr. Andrew Regenscheid
andrew.regenscheid@nist.gov

SP 800-146, *Cloud Computing Synopsis and Recommendations*

This document reprises the NIST-established definition of cloud computing, describes cloud computing benefits and open issues, presents an overview of major classes of cloud technology, and provides guidelines and recommendations on how organizations should consider the relative opportunities and risks of cloud computing. Cloud computing has been the subject of a great deal of commentary. Attempts to describe cloud computing in general terms, however, have been problematic because cloud computing is not a single kind of system, but instead spans a spectrum of underlying technologies, configuration possibilities, service models, and deployment models. This document describes cloud systems and discusses their strengths and weaknesses.

Contacts:

Mr. Mark Lee Badger   Mr. Tim Grance
mark.badger@nist.gov   grance@nist.gov

Mr. Jeff Voas
jeff.voas@nist.gov

SP 800-144, *Guidelines on Security and Privacy in Public Cloud Computing*

Cloud computing can and does mean different things to different people. The common characteristics most interpretations share are on-demand scalability of highly available and reliable pooled computing resources, secure access to metered services from nearly anywhere, and displacement of data and services from inside to outside the organization. While aspects of these characteristics have been realized to a certain extent, cloud computing remains a work in progress. This publication provides an overview of the security and privacy challenges pertinent to public cloud computing and points out considerations organizations should take when outsourcing data, applications, and infrastructure to a public cloud environment.

Contact:

Mr. Tim Grance
grance@nist.gov

SP 800-135 Rev. 1, *Recommendation for Existing Application-Specific Key Derivation Functions*

Cryptographic keys are vital to the security of Internet security applications and protocols. Many widely used Internet security protocols have their own application-specific Key Derivation Functions (KDFs) that are used to generate the cryptographic keys required for their cryptographic functions. This recommendation provides security requirements for those KDFs.

Contact:

Mr. Quynh Dang
quynh.dang@nist.gov

Draft SP 800-130, *A Framework for Designing Cryptographic Key Management Systems*

This Framework for Designing Cryptographic Key Management Systems (CKMS) contains topics that should be considered by a CKMS designer when developing a CKMS design specification. For each topic, there are one or more documentation requirements that need to be addressed by the design specification. Thus, any CKMS that adequately addresses these requirements would have a design specification that is compliant with this Framework.

Contact:

Ms. Elaine Barker
elaine.barker@nist.gov

Draft SP 800-124 Rev. 1, *Guidelines for Managing and Securing Mobile Devices in the Enterprise*

Mobile devices, such as smart phones and tablets, typically need to support multiple security objectives: confidentiality, integrity, and availability. To achieve these objectives, mobile devices should be secured against a variety of threats. The purpose of this publication is to help organizations centrally manage and secure mobile devices. Laptops are out of the scope of this publication, as are mobile devices with minimal computing capability, such as basic cell phones. This publication provides recommendations for selecting, implementing, and using centralized management technologies, and it explains the security concerns inherent in mobile device use and provides recommendations for securing mobile devices throughout their life cycles. The scope of this publication includes securing both organization-provided and personally owned (bring your own device) mobile devices.

Contact:
Mr. Murugiah Souppaya
murugiah.souppaya@nist.gov

SP 800-121 Rev. 1, *Guide to Bluetooth Security*

Bluetooth is an open standard for short-range radio frequency communication. Bluetooth technology is used primarily to establish wireless personal area networks (WPANs), and it has been integrated into many types of business and consumer devices. This publication provides information on the security capabilities of Bluetooth technologies and gives recommendations to organizations employing Bluetooth technologies on securing them effectively. The Bluetooth versions within the scope of this publication are versions 1.1, 1.2, 2.0 + Enhanced Data Rate (EDR), 2.1 + EDR, 3.0 + High Speed (HS), and 4.0, which includes Low Energy (LE) technology.

Contact:
Dr. Lily Chen
lily.chen@nist.gov

Draft SP 800-117 Rev. 1, *Guide to Adopting and Using the Security Content Automation Protocol (SCAP) Version 1.2*

The purpose of this document is to provide an overview of the Security Content Automation Protocol (SCAP) version 1.2. This document discusses SCAP at a conceptual level, focusing on how organizations can use SCAP-enabled tools to enhance their security posture. It also explains to IT product and service vendors how they can adopt SCAP version 1.2 capabilities within their offerings. The intended audience for this

document is individuals who have responsibilities for maintaining or verifying the security of systems in operational environments.

Contacts:
Mr. Stephen Quinn
stephen.quinn@nist.gov

Mr. David Waltermire
david.waltermire@nist.gov

SP 800-107 Rev. 1, *Recommendation for Applications Using Approved Hash Algorithms*

Hash functions that compute a fixed-length message digest from arbitrary length messages are widely used for many purposes in information security. This document provides security guidelines for achieving the required or desired security strengths when using cryptographic applications that employ the approved hash functions specified in Federal Information Processing Standard (FIPS) 180-4, *Secure Hash Standard (SHS)*. These include functions such as digital signatures, Keyed-hash Message Authentication Codes (HMACs), and Hash-based Key Derivation Functions (Hash-based KDFs).

Contact:
Mr. Quynh Dang
quynh.dang@nist.gov

Draft SP 800-94 Rev. 1, *Guide to Intrusion Detection and Prevention Systems (IDPSs)*

Intrusion detection and prevention systems (IDPSs) are focused on identifying possible incidents, logging information about them, attempting to stop them, and reporting them to security administrators. In addition, organizations use IDPSs for other purposes, such as identifying problems with security policies, documenting existing threats, and deterring individuals from violating security policies. This publication describes the characteristics of IDPS technologies and provides recommendations for designing, implementing, configuring, securing, monitoring, and maintaining them. The types of IDPS technologies are differentiated primarily by the types of events that they monitor and the ways in which they are deployed. This publication discusses the following four types of

IDPS technologies: network-based, wireless, network behavior analysis (NBA), and host-based.

Contact:
Mr. Peter Mell
mell@nist.gov

Draft SP 800-90C, *Recommendation for Random Bit Generator (RBG) Constructions*

This recommendation specifies constructions for the implementation of random bit generators (RBGs). An RBG may be a deterministic random bit generator (DRBG) or a non-deterministic random bit generator (NRBG). The constructed RBGs consist of DRBG mechanisms as specified SP 800-90A, *Recommendation for Random Number Generation Using Deterministic Random Bit Generators*, and entropy sources as specified in (draft) SP 800-90B, *Recommendation for the Entropy Sources Used for Random Bit Generation.*

Contacts:
Ms. Elaine Barker   Dr. John Kelsey
elaine.barker@nist.gov   john.kelsey@nist.gov

Draft SP 800-90B, *Recommendation for the Entropy Sources Used for Random Bit Generation*

This recommendation specifies the design principles and requirements for the entropy sources used by Random Bit Generators, and the tests for the validation of entropy sources. These entropy sources are intended to be combined with Deterministic Random Bit Generator mechanisms that are specified in SP 800-90A, *Recommendation for Random Number Generation Using Deterministic Random Bit Generators*, to construct Random Bit Generators, as specified in (draft) SP 800-90C, *Recommendation for Random Bit Generator (RBG) Constructions.*

Contacts:
Ms. Elaine Barker   Dr. John Kelsey
elaine.barker@nist.gov   john.kelsey@nist.gov

SP 800-90 A, *Recommendation for Random Number Generation Using Deterministic Random Bit Generators*

This recommendation specifies mechanisms for the generation of random bits using deterministic methods. The methods provided are based on hash functions, block cipher algorithms, or number theoretic problems.

Contacts:
Ms. Elaine Barker   Dr. John Kelsey
elaine.barker@nist.gov   john.kelsey@nist.gov

Draft SP 800-88 Rev. 1, *Guidelines for Media Sanitization*

This document will assist organizations in implementing a media sanitization program with proper and applicable techniques and controls for sanitization and disposal decisions, considering the security categorization of the associated system's confidentiality. The objective of this special publication is to assist with decision making when media require disposal, reuse, or will be leaving the effective control of an organization. Organizations should develop and use local policies and procedures in conjunction with this guide to make effective, risk-based decisions on the ultimate sanitization and/or disposition of media and information.

Contacts:
Mr. Richard Kissel   Mr. Matthew Scholl
richard.kissel@nist.gov   matthew.scholl@nist.gov

Draft SP 800-83 Rev. 1, *Guide to Malware Incident Prevention and Handling for Desktops and Laptops*

Malware, also known as malicious code, refers to a program that is covertly inserted into another program with the intent to destroy data, run destructive or intrusive programs, or otherwise compromise the confidentiality, integrity, or availability of the victim's data, applications, or operating system. Malware is the most common external threat to most hosts, causing widespread damage and disruption and necessitating extensive recovery efforts within most organizations. This publication provides recommendations for improving an organization's malware incident prevention measures. It also gives extensive recommendations for enhancing an organization's existing incident response capability so that it is better prepared to handle malware incidents, particularly widespread ones.

Contact:
Mr. Murugiah Souppaya
murugiah.souppaya@nist.gov

Draft SP 800-76-2, *Biometric Data Specification for Personal Identity Verification*

Homeland Security Presidential Directive (HSPD-12) called for new standards to be adopted governing interoperable use of identity credentials to allow physical and logical access to federal government locations and systems. The Personal Identity Verification (PIV) standard for Federal Employees and Contractors, FIPS 201, was developed to define procedures and specifications for issuance and use of an interoperable identity credential. This document is a companion document to FIPS 201 and describes technical acquisition and formatting specifications for the PIV system, including the PIV Card itself. It also establishes minimum accuracy specifications for deployed biometric authentication processes. The approach is to enumerate procedures and formats for collection and preparation of fingerprint, iris and facial data, and to restrict values and practices included generically in published biometric standards. The primary design objective behind these particular specifications is high performance and universal interoperability. The addition of iris and face specifications in the 2012 edition adds an alternative modality for biometric authentication and extends coverage to persons for whom fingerprinting is problematic. The addition of on-card comparison offers an alternative to PIN-mediated card activation as well as an additional authentication method. For the preparation of biometric data suitable for the Federal Bureau of Investigation (FBI) background check, SP 800-76 references FBI documentation, including the ANSI/NIST Fingerprint Standard and the Electronic Fingerprint Transmission Specification. This document does not preclude use of other biometric modalities in conjunction with the PIV card.

Contact:
Mr. Patrick Grother
patrick.grother@nist.gov

SP 800-67 Rev. 1, *Recommendation for the Triple Data Encryption Algorithm (TDEA) Block Cipher*

This publication specifies the Triple Data Encryption Algorithm (TDEA), including its primary component cryptographic engine, the Data Encryption Algorithm (DEA). When implemented in an SP 800-38 series-compliant mode of operation and in a FIPS 140-2-compliant cryptographic module, TDEA may be used by federal organizations to protect sensitive unclassified data. Protection of data during transmission or while in storage may be necessary to maintain the confidentiality and integrity of the information represented by the data. This publication defines the mathematical steps required to cryptographically protect data using TDEA and to subsequently process such protected data. TDEA is made available for use by federal agencies within the context of a total security program consisting of physical security procedures, good information management practices, and computer system/network access controls.

Contact:
Ms. Elaine Barker
elaine.barker@nist.gov

SP 800-63-1, *Electronic Authentication Guideline*

This recommendation provides technical guidelines for federal agencies implementing electronic authentication and is not intended to constrain the development or use of standards outside of this purpose. The recommendation covers remote authentication of users (such as employees, contractors, or private individuals) interacting with government IT systems over open networks. It defines technical requirements for each of four levels of assurance in the areas of identity proofing, registration, tokens, management processes, authentication protocols, and related assertions.

Contacts:

Dr. Lily Chen
lily.chen@nist.gov

Dr. Ray Perlner
ray.perlner@nist.gov

Ms. Donna Dodson
donna.dodson@nist.gov

Mr. William Polk
william.polk@nist.gov

SP 800-61 Rev. 2, *Computer Security Incident Handling Guide*

Computer security incident response has become an important component of information technology programs. Because performing incident response effectively is a complex undertaking, establishing a successful incident response capability requires substantial planning and resources. This publication assists organizations in establishing computer security incident response capabilities and handling incidents efficiently and effectively. This publication provides guidelines for incident handling, particularly for analyzing incident-related data and determining the appropriate response to each incident. The guidelines can be followed independently of particular hardware platforms, operating systems, protocols, or applications.

Contact:
Mr. Tim Grance
grance@nist.gov

SP 800-57 Part 1, *Recommendation for Key Management: Part 1: General (Revision 3)*

This recommendation provides cryptographic key management guidance. It consists of three parts. Part 1 provides general guidance and best practices for the management of cryptographic keying material. Part 2 provides guidance on policy and security planning requirements for U.S. government agencies. Finally, Part 3 provides guidance when using the cryptographic

features of current systems.

Contacts:

Ms. Elaine Barker
elaine.barker@nist.gov

Mr. William Polk
william.polk@nist.gov

SP 800-56 C, *Recommendation for Key Derivation through Extraction-then-Expansion*

This recommendation specifies techniques for the derivation of keying material from a shared secret established during a key establishment scheme defined in Special Publications 800-56A, *Recommendation for Pair-Wise Key Establishment Schemes Using Discrete Logarithm Cryptography (Revised)*, or 800-56B, *Recommendation for Pair-Wise Key Establishment Schemes Using Integer Factorization Cryptography*, through an extraction-then-expansion procedure.

Contact:
Dr. Lily Chen
lily.chen@nist.gov

Draft SP 800-56A Rev. 1, *Recommendation for Pair-Wise Key-Establishment Schemes Using Discrete Logarithm Cryptography*

This recommendation specifies key-establishment schemes based on the discrete logarithm problem over finite fields and elliptic curves, including several variations of Diffie-Hellman and Menezes-Qu-Vanstone (MQV) key establishment schemes.

Contacts:

Ms. Elaine Barker
elaine.barker@nist.gov

Dr. Lily Chen
lily.chen@nist.gov

Dr. Allen Roginsky
allen.roginsky@nist.gov

Draft SP 800-53 Rev. 4, *Security and Privacy Controls for Federal Information Systems and Organizations*

Special Publication (SP) 800-53, Revision 4, represents the culmination of a year-long initiative to update the content of the security controls catalog and the guidance for selecting and specifying security controls for federal information systems and organizations. The project was conducted as part of the Joint Task Force Transformation Initiative in cooperation and collaboration with the Department of Defense, the Intelligence Community, the Committee on National Security Systems, and the Department of Homeland Security. The proposed changes included in Revision 4 are directly linked to the current state of the threat space (i.e., capabilities, intentions, and targeting activities of adversaries) and the attack data collected and analyzed over a substantial time period. In particular, the major changes in Revision 4 include:

* New security controls and control enhancements;
* Clarification of security control requirements and specification language;
* New tailoring guidance including the introduction of overlays;
* Additional supplemental guidance for security controls and enhancements;
* New privacy controls and implementation guidance;
* Updated security control baselines;
* New summary tables for security controls to facilitate ease-of-use; and
* Revised minimum assurance requirements and designated assurance controls.

Contacts:
Dr. Ron Ross   NIST FISMA Team
rross@nist.gov   sec-cert@nist.gov

Draft SP 800-40 Rev. 3, *Guide to Enterprise Patch Management Technologies*

Patch management is the process for identifying, acquiring, installing, and verifying patches for products and systems. Patches correct security and functionality problems in software and firmware. There are several challenges that complicate patch management. If organizations do not overcome these challenges, they will be unable to patch systems effectively and efficiently, leading to easily preventable compromises.

This publication is designed to assist organizations in understanding the basics of enterprise patch management technologies. It explains the importance of patch management and examines the challenges inherent in performing patch management. It provides an overview of enterprise patch management technologies, and it also briefly discusses metrics for measuring the technologies' effectiveness and for comparing the relative importance of patches.

Contact:
Mr. Murugiah Souppaya
murugiah.souppaya@nist.gov

SP 800-30 Rev. 1, *Guide for Conducting Risk Assessments*

The purpose of this publication is to provide guidance for conducting risk assessments of federal information systems and organizations, amplifying the guidance provided in Special Publication 800-39, *Managing Information Security Risk: Organization, Mission, and Information System View*. This document provides guidance for carrying out each of the three steps in the risk assessment process (i.e., prepare for the assessment, conduct the assessment, and maintain the assessment) and how risk assessments and other organizational risk management processes complement and inform each other.

Contacts:
Dr. Ron Ross   NIST FISMA Team
rross@nist.gov   sec-cert@nist.gov

→ NIST Interagency Reports

NISTIR 7877, *BioCTS 2012: Advanced Conformance Test Architectures and Test Suites for Biometric Data Interchange Formats and Biometric Information Records*

BioCTS 2012 is biometric conformance test software designed to test implementations for conformance to various biometric data interchange format standards.

BioCTS 2012 for the American National Standards Institute (ANSI)/NIST-ITL 1-2011 tests implementations of NIST SP 500-290, *ANSI/NIST ITL 1-2011 (AN-2011) Data Format for the Interchange of Fingerprint, Facial & Other Biometric Information*, using test assertions documented in NIST SP 500-295, *Conformance Testing Methodology for ANSI/NIST-ITL 1-2011, Data Format for the Interchange of Fingerprint, Facial & Other Biometric Information (Release 1.0)*. BioCTS 2012 for the International Organization for Standardization/ International Electrotechnical Commission (ISO/IEC) tests implementations of biometric data interchange formats developed by Subcommittee 37 - Biometrics of the Joint Technical Committee 1 - Information Technology of ISO and IEC. Support for testing Biometric Information Records (BIRs) conforming to instantiations of the Common Biometric Exchange Formats Framework (CBEFF) specified in national and international standards is also provided. BioCTS 2012 for ANSI/NIST-ITL 1-2011 is currently designed to support testing of implementations that include any of the Record Types defined in AN-2011, but conformance testing is only performed for the selected Record Types (1, 4, 10, 13, 14, 15, and 17). Plans exist to extend the test tool to support additional Record Types. Information regarding BioCTS 2012 testing architectures, code structure, and other software design details is provided.

Contacts:

Mr. Fernando Podio   Mr. Dylan Yaga
fernando.podio@nist.gov   dylan.yaga@nist.gov

NISTIR 7874, *Guidelines for Access Control System Evaluation Metrics*

Nearly all applications include some form of access control (AC). AC is concerned with determining the allowed activities of legitimate users, mediating every attempt by a user to access a resource in the system. AC systems come with a wide variety of features and administrative capabilities, and their operational impact can be significant. In particular, this impact can pertain to administrative and user productivity, as well as to the organization's ability to perform its mission. Therefore, it is reasonable to use quality

metrics to verify the mechanical properties of AC systems. This document discusses the administration, enforcement, performance, and support properties of AC mechanisms that are embedded in each AC system. Because of the rigorous nature of the metrics and the knowledge needed to gather them, these metrics are intended to be used by AC experts who are evaluating the highest-security AC systems.

Contact:

Dr. Vincent Hu
vhu@nist.gov

NISTIR 7870, *NIST Test Personal Identity Verification (PIV) Cards*

In order to facilitate the development of applications and middleware that support the Personal Identity Verification (PIV) Card, NIST has developed a set of test PIV Cards and a supporting public key infrastructure. This set of test cards includes not only examples that are similar to cards issued today, but also examples of cards with features that are expected to appear in cards that will be issued in the future. This document provides an overview of the test cards and the infrastructure that has been developed to support their use.

Contact:

Dr. David Cooper
david.cooper@nist.gov

NISTIR 7864, *The Common Misuse Scoring System (CMSS): Metrics for Software Feature Misuse Vulnerabilities*

The Common Misuse Scoring System (CMSS) is a set of measures of the severity of software feature misuse vulnerabilities. A software feature is a functional capability provided by software. A software feature misuse vulnerability is a vulnerability in which the feature also provides an avenue to compromise the security of a system. Such vulnerabilities are present when the trust assumptions made when designing software features can be abused in ways that violate security. Misuse vulnerabilities allow attackers to use for malicious purposes the functionality that was intended to be beneficial. CMSS can provide measurement data to assist organizations in making sound decisions on addressing software feature misuse vulnerabilities and in conducting quantitative assessments of the overall security posture of a system. This report defines proposed measures for CMSS and equations to be used to combine the measures into severity scores for each vulnerability. The report also provides examples of how CMSS measures and scores would be determined for selected software feature misuse vulnerabilities.

Contact:
Mr. Peter Mell
mell@nist.gov

Draft NISTIR 7848, *Specification for the Asset Summary Reporting Format 1.0*

This specification defines the Asset Summary Reporting (ASR) format version 1.0, a data model for expressing the data exchange format of summary information relative to one or more metrics. ASR reduces the bandwidth requirement to report information about assets in the aggregate since it allows for reporting aggregates relative to metrics, as opposed to reporting data about each individual asset, which can lead to a bloated data exchange. ASR is vendor-neutral and leverages widely adopted, open specifications; it is flexible, and suited for a wide variety of reporting applications.

Contact:
Mr. David Waltermire
david.waltermire@nist.gov

Draft NISTIR 7831, *Common Remediation Enumeration (CRE) Version 1.0*

This document defines the Common Remediation Enumeration (CRE) 1.0 specification. CRE is part of a suite of enterprise remediation specifications that enable automation and enhanced correlation of enterprise remediation activities. Each CRE entry represents a unique remediation activity and is assigned a globally unique CRE identifier (CRE-ID). This specification describes the core concepts of CRE, the

technical components of a CRE entry, outlines how CRE entries are created, the technical requirements for constructing a CRE-ID, and how CRE-IDs may be assigned. CRE-IDs are intended to be boundary objects that are broadly useable in enterprise security management products and information domains that participate in remediation activities or make assertions about remediation actions.

Contact:
Mr. David Waltermire
david.waltermire@nist.gov

Draft NISTIR 7823, *Advanced Metering Infrastructure Smart Meter Upgradeability Test Framework*

As electric utilities turn to Advanced Metering Infrastructures (AMIs) to promote the development and deployment of the Smart Grid, one aspect that can benefit from standardization is the upgradeability of Smart Meters. The National Electrical Manufacturers Association (NEMA) standard SG-AMI 1-2009, *Requirements for Smart Meter Upgradeability*, describes functional and security requirements for the secure upgrade—both local and remote—of Smart Meters. This report describes conformance test requirements that may be used voluntarily by testers and/or test laboratories to determine whether Smart Meters and Upgrade Management Systems conform to the requirements of NEMA SG-AMI 1-2009. For each relevant requirement in NEMA SG-AMI 1-2009, the document identifies the information to be provided by the vendor to facilitate testing, and the high-level test procedures to be conducted by the tester/laboratory to determine conformance.

Contact:
Dr. Michaela Iorga
michaela.iorga@nist.gov

Draft NISTIR 7817, *A Credential Reliability and Revocation Model for Federated Identities*

A large number of IDentity Management Systems (IDMSs) are being deployed worldwide that use different technologies for the population of their users. With the diverse set of technologies, and the unique business requirements for organizations to federate, there is no uniform approach to the federation process. Similarly, there is no uniform method to revoke credentials or their associated attribute(s) in a federated community. In the absence of a uniform revocation method, this document seeks to investigate credential and attribute revocation with a particular focus on identifying missing requirements for credential and attribute revocation. This document first introduces and analysis the different types of digital credentials and recommends missing revocation-related requirements for each model in a federated environment. As a second goal, and as by-product of the analysis and recommendations, this paper suggests a credential reliability and revocation service that serves to eliminate the missing requirements and involves all the entities of the federation.

Contact:
Ms. Hildegard Ferraiolo
hildegard.ferraiolo@nist.gov

NISTIR 7816, *2011 Computer Security Division Annual Report*

Title III of the E-Government Act of 2002, *Federal Information Security Management Act (FISMA)* of 2002, requires NIST to prepare an annual public report on activities undertaken in the previous year, and planned for the coming year, to carry out responsibilities under this law. The primary goal of the NIST's Computer Security Division (CSD) is to provide standards and technology that protects information systems against threats to the confidentiality, integrity, and availability of information and services. During Fiscal Year 2011 (FY2011), CSD successfully responded to numerous challenges and opportunities in fulfilling that mission. Through CSD's diverse research agenda and engagement in many national priority initiatives, high-quality,

cost-effective security and privacy mechanisms were developed and applied that improved information security across the federal government and the greater information security community. This annual report highlights the research agenda and activities in which CSD was engaged during FY2011.

Contacts:
Mr. Patrick O'Reilly   Mr. Kevin Stine
patrick.oreilly@nist.gov   kevin.stine@nist.gov

Draft NISTIR 7800, *Applying the Continuous Monitoring (CM) Technical Reference Model to the Asset, Configuration, and Vulnerability Management Domains*

This publication binds together the CM workflows and capabilities described in draft NISTIR 7799, *Continuous Monitoring Reference Model Workflow, Subsystem, and Interface Specifications*, to specific data domains. It focuses on the Asset Management, Configuration, and Vulnerability data domains. It leverages the Security Content Automation Protocol (SCAP) version 1.2 for configuration and vulnerability scan content, and it dictates reporting results in an SCAP-compliant format. This specification describes an overview of the approach to each of the three domains, how they bind to specific communication protocols, and how those protocols interact. It then defines the specific requirements levied upon the various capabilities of the subsystems defined in draft NISTIR 7799 that enable each data domain.

Contacts:
Mr. David Waltermire   Mr. Peter Mell
david.waltermire@nist.gov   mell@nist.gov

Draft NISTIR 7799, *Continuous Monitoring Reference Model Workflow, Subsystem, and Interface Specifications*

This publication provides the technical specifications for the continuous monitoring (CM) reference model presented in draft NISTIR 7756, *CAESARS Framework Extension: An Enterprise Continuous Monitoring Technical Reference Architecture*. These specifications enable multi-instance CM implementations, hierarchical tiers, multi-instance dynamic querying, sensor tasking, propagation of policy, policy monitoring, and policy compliance reporting. A major focus of the specifications is on workflows that describe the coordinated operation of all subsystems and components within the model. Another focus is on subsystem specifications that enable each subsystem to play its role within the workflows. The final focus is on interface specifications that supply communication paths between subsystems. These three sets of specifications (workflows, subsystems, and interfaces) are written to be data domain-agnostic, which means that they can be used for CM regardless of the data domain that is being monitored. A companion publication, draft NIST IR 7800, *Applying the Continuous Monitoring Technical Reference Model to the Asset, Configuration, and Vulnerability Management Domains*, binds these specifications to specific data domains (e.g., asset, configuration, and vulnerability management). The specifications provided in this document are detailed enough to enable product instrumentation and development. They are also detailed enough to enable product testing, validation, procurement, and interoperability. Taken together, the specifications in this document define an ecosystem where a variety of interoperable products can be composed together to form effective CM solutions. If properly adopted, these specifications will enable teamwork, orchestration, and coordination among CM products that currently operate distinctly. For the computer security domain, this will greatly enhance organizational effectiveness and efficiency in addressing known vulnerabilities and technical policy requirements, and decision making.

Contacts:
Mr. David Waltermire   Mr. Peter Mell
david.waltermire@nist.gov   mell@nist.gov

Draft NISTIR 7756, *CAESARS Framework Extension: An Enterprise Continuous Monitoring Technical Reference Architecture*

This publication and its supporting documents present an enterprise continuous monitoring technical reference model that extends the framework provided by the DHS Federal Network Security CAESARS architecture. This extension enables added functionality, defines each subsystem in more detail, and further leverages security automation standards. It also extends CAESARS to allow for large implementations that need a multi-tier architecture and focuses on the necessary inter-tier communications. The goal of this document is to facilitate enterprise continuous monitoring by presenting a reference model that enables organizations to aggregate collected data from across a diverse set of security tools, analyze that data, perform scoring, enable user queries, and provide overall situational awareness. The model design is focused on enabling organizations to realize this capability by leveraging their existing security tools and thus avoiding complicated and resource-intensive custom tool integration efforts.

Contacts:
Mr. David Waltermire   Mr. Peter Mell
david.waltermire@nist.gov   mell@nist.gov

Mr. Harold Booth
harold.booth@nist.gov

Draft NISTIR 7622, *Notional Supply Chain Risk Management Practices for Federal Information Systems*

This publication is intended to provide a wide array of practices that, when implemented, will help mitigate supply chain risk to federal information systems. It seeks to equip federal departments and agencies with a notional set of repeatable and commercially reasonable supply chain assurance methods and practices that offer a means to obtain an understanding of, and visibility throughout, the supply chain.

Contacts:
Mr. Jon Boyen   Ms. Celia Paulsen
jon.boyens@nist.gov   celia.paulsen@nist.gov

Draft NISTIR 7511 Rev. 3.04, *Security Content Automation Protocol (SCAP) Version 1.0 Validation Program Test Requirements*

This publication defines the requirements and associated test procedures necessary for products to achieve one or more Security Content Automation Protocol (SCAP) validations. Validation is awarded based on a defined set of SCAP capabilities by independent laboratories that have been accredited for SCAP testing by the National Voluntary Laboratory Accreditation Program (NVLAP).

Contacts:

Mr. David Waltermire
david.waltermire@nist.gov

Mr. Stephen Quinn
stephen.quinn@nist.gov

Ms. Melanie Cook
melanie.cook@nist.gov

Mr. John Banghart
john.banghart@nist.gov

Guest Research Internships at NIST

Opportunities are available at NIST for 6- to 24-month internships within CSD. Qualified individuals should contact CSD, provide a statement of qualifications, and indicate the area of work that is of interest. Generally speaking, the salary costs are borne by the sponsoring institution; however, in some cases, these guest research internships carry a small monthly stipend paid by NIST. For further information, contact:

Ms. Donna Dodson
(301) 975-8443
donna.dodson@nist.gov

Mr. Matthew Scholl
(301) 975-2941
matthew.scholl@nist.gov

Details at NIST for Government or Military Personnel

Opportunities are available at NIST for 6- to 24-month details at NIST in CSD. Qualified individuals should contact CSD, provide a statement of qualifications, and indicate the area of work that is of interest. Generally speaking, the salary costs are borne by the sponsoring agency; however, in some cases, agency salary costs may be reimbursed by NIST. For further information, contact:

Ms. Donna Dodson
(301) 975-8443
donna.dodson@nist.gov

Mr. Matthew Scholl
(301) 975-2941
matthew.scholl@nist.gov

Federal Computer Security Program Managers' Forum (FCSPM)

The FCSPM Forum is covered in detail in the Outreach section of this report. Membership is free and open to federal employees. For further information, contact:

Mr. Kevin Stine
(301) 975-4483
kevin.stine@nist.gov or sec-forum@nist.gov
visit the FCSPM Forum website at
http://csrc.nist.gov/groups/SMA/
forum/membership.html

Security Research

NIST occasionally undertakes security work, primarily in the area of research, funded by other agencies. Such sponsored work is accepted by NIST when it can cost-effectively further the goals of NIST and the sponsoring institution. For further information, contact:

Ms. Donna Dodson
(301) 975-8443
donna.dodson@nist.gov

Funding Opportunities at NIST

NIST funds industrial and academic research in a variety of ways. The Small Business Innovation Research Program funds R&D proposals from small businesses; see www.nist.gov/sbir. CSD also offers other grants to encourage work in specific fields: precision measurement, fire research, and materials science. Grants/awards supporting research at industry, academia, and other institutions are available on a competitive basis through several different Institute offices.

For general information on NIST grants programs, please contact:

Mr. Christopher Hunton
(301) 975-5718
christopher.hunton@nist.gov

Further details on funding opportunities may be found on http://www.nist.gov/director/ocfo/grants/grants.cfm

The editor, Patrick O'Reilly of the Computer Security Division, wishes to thank his colleagues in the Computer Security Division, who provided write-ups on their 2012 project highlights and accomplishments for this annual report (their names are mentioned after each project write-up). The editor would also like to acknowledge Elizabeth Lennon, Lisa Carnahan, Kevin Stine, Jim Foti, and Peggy Himes for reviewing and providing valuable feedback for this annual report.

www.ingramcontent.com/pod-product-compliance
Lightning Source LLC
Chambersburg PA
CBHW081550170526
45166CB00009B/2648